EARTH RESOURCES RESEARCH

Earth Resources Research was established in 1973 to investigate contemporary environmental and resource management issues. It is a small versatile research group, specialising in the fields of energy, agriculture, transport and wildlife. Registered as a charity, it is an independent organisation with a qualified staff backed up by outside specialists. The aim is to provide thorough, challenging research directly applicable to current policy.

This report is one of a series concerned with food and agricultural policy. Details of other publications in print appear on the inside back cover. Requests for further information about the organisation and enquiries about this report should be sent to the address below. Comments and suggestions are always welcome.

David Baldock
Executive Director
Earth Resources Research
258 Pentonville Road
London N1 9JX

ACKNOWLEDGEMENTS

I am grateful to many people for help with my work, especially the farmers who provided me with information and showed me around their farms. Bernard Peet, Geoff Hearnden, Roger Ewbank, Pete Ffitch, Joanne Bower, Geoff Cloke and Sue Leejeffs all commented upon the first draft. Ruth Harrison and David Sainsbury gave encouragement. Maria Rosenthal accompanied me on a number of farm visits. Caren Levy and Clare Cherrington helped edit the final draft. I am particularly indebted to David Baldock for all his advice and support for editing the various drafts through which this study progressed.

The views expressed in this book are my own and not necessarily those of the sponsoring organisations. Notwithstanding all the assistance I received, responsibility for any errors rests with me.

Alternatives to Factory Farming

Paul Carnell

First published 1983

Earth Resources Research Ltd
258 Pentonville Road
London N1 9JY

© 1983 Earth Resources Research Ltd

Typeset and printed by
Modern Text Typesetting

Cover design by Reg. Boorer,
Graphics by Barry Savage

Carnell, Paul
    Alternatives to factory farming.
    1.   Animal industry
    1.   Title
    636   HD9415

ISBN 0-946281-01-7

Contents

Figures

Tables

Preface

"Factory Farming" is a graphic term for modern systems of intensive husbandry which have largely removed pigs and poultry from the rural landscape and confined them in buildings. Over the last thirty years intensive systems have achieved a rapid pre-eminence as traditional forms of stock-rearing have fallen into disuse. It is the alternatives which are now unorthodox.

In the haste to exploit the revolution in stock-rearing, fundamental questions of animal welfare were left unanswered and the great bulk of research focussed on improvements in efficiency. It is only comparatively recently that public unease has forced a serious reappraisal of welfare issues, revealing considerable gaps in our knowledge and substantive disagreement about the acceptability of certain methods. Despite the scepticism of the industry, alternative systems are now receiving close attention and many of them appear to offer major advantages from a welfare point of view. Their commercial acceptability is another question. It has been widely argued by both government and industry sources that consumers would have to pay considerably higher prices if factory farming techniques were abandoned. The report that follows arose from doubts about the basis of some of these claims, and a concern that they were being made by bodies with a predisposition to judge the alternatives rather harshly. In our view, a thorough, independent study was required.

Paul Carnell's study does not attempt to consider all the possible alternatives to "factory farming". It is concerned solely with egg production and the breeding stage of pig production. The report is a guide to alternatives currently in use and an examination of how they might be expected to perform under commercial conditions. It is primarily an economic analysis and the results are expressed in the form of costings, thus providing a comparison with contemporary intensive systems. The report looks also at some of the broader issues involved, such as environmental pollution, but there is no attempt to quantify these as costs and benefits. Nor have we tried to assess the welfare aspects of different systems—although this is an urgent task, it requires rather different skills.

Our hope is that this report will challenge some ill-founded assumptions about alternative systems and stimulate a more rigorous discussion about certain key economic issues. Animal welfare is inevitably an emotional issue, but clarity is essential in judging the economic aspects of the debate. By its very nature, part of this report is theoretical, but it also contains a good deal of practical information and we hope that it will be of some value to farmers and others in the livestock industry. Comments from any quarter will certainly be welcome.

Independent studies of this kind are frequently difficult to fund and we are grateful to our sponsors, in particular the Universities Federation for Animal Welfare. Without the Federation's support it would not have been possible to complete the necessary research or publish the results. Preliminary findings were reported at the Federation's symposium on farm animal welfare held in 1981 and also submitted to the Parliamentary Select Committee on Agriculture which reported on the same topic later in the year. We would also like to thank the Farm and Food Society, Education Services Ltd. and the Augustine Trust, all of which helped to fund the early stages of the project, and the St Andrew Animal Fund which made a grant towards the publication of the final report. Needless to say, none of these bodies is in any way responsible for the views expressed herein.

David Baldock
Executive Director

1 Introduction

1. Background to the study

Changes in livestock husbandry since the war have made farm animal welfare a controversial issue. Recent demonstrations of public concern have prompted government scrutiny of the subject, both in the UK and elsewhere in Europe. In West Germany, legal proceedings taken by a welfare group against a battery cage egg producer led to a judgement which threatened the legality of the cage system and forced the attention of the European Commission. In July 1981 in the UK, the Agriculture Select Committee of the House of Commons reported on animal welfare in poultry, pig and veal production. Their report accepted the case for some major changes in livestock production methods to improve animal welfare (although by a narrow majority in some cases), including the abolition of the battery cage and the provision of bedded areas for all pigs housed indoors.

Higher stocking densities, and more closely controlled and less 'natural' environments underlie much of the concern expressed by welfare groups. Close confinement of stock, declining use of straw and lack of access to outdoors have provoked considerable criticism. Attention has focussed most sharply on battery cage egg production and the use of veal crates, but there is pressure to ban stall and tether housing for sows and to limit the early-weaning of pigs. There is also anxiety that cattle may follow the path of pigs and poultry if 'zero grazing' and 'barley beef' are seen to have significant commercial attractions.

Farmers and their representatives have reacted strongly to criticism, rejecting most objections to intensive farming as emotive and unrealistic. In general, intensive methods are justified on the grounds that they permit a more efficient use of resources, particularly labour and feedstuffs. Furthermore, farmers' representatives are strongly opposed to any regulation of their activities.

Controversy has stimulated research into the welfare implications of intensification and there has been some effort to develop alternative systems. However, there has been virtually no detailed study of economic aspects. This was clear from the dearth of evidence on this area submitted to the Agriculture Select Committee. Of course, economic issues are not necessarily of most importance when considering

animal welfare. The Select Committee was emphatic on this point:
'We do not accept the contention, frequently stated or implied, that the public demand for cheap food decrees that the cheapest possible methods of production must be adopted . . . society has the duty to see that undue suffering is not caused to animals, and we cannot accept that that duty should be set aside in order that food may be produced more cheaply. Where unacceptable suffering can be eliminated only at extra cost, that cost should be borne or the product foregone.'[1]

This does not, however, mean that economic issues are unimportant. It is clearly useful to know how various welfare regulations might affect costs and resource use. Furthermore, Governments do tend in practice to give considerable weight to economic considerations, whether rightly or wrongly, as the White Paper responding to the Select Committee's report made clear:

'The Government believes that, whilst the cheapest methods of production may not always be the most appropriate, the likely effects on prices to the consumer must be taken into account both in setting welfare standards and in the timing of any change in standards.'[2]

It is important, therefore, that policy makers are well-informed on this issue, particularly as ignorance can be invoked to justify inertia.

We undertook the present study in an attempt to provide some detailed and carefully supported comparisons of the economic performance of different systems. Our budget, and the amount of effort needed to glean and synthesise information from highly diverse sources, quickly forced us to narrow the scope of the project so as to be able to retain the depth which we sought. We focussed, therefore, mainly upon the 'on farm' costs—the costs of production to the farmer—and restricted our attention to breeding pigs* and egg production.

This approach has inevitably limited the scope of our work, although later in the book there is a discussion of some of the wider costs of livestock production which need to be considered in a full economic appraisal.

The next section briefly examines the development of intensive livestock production, and this chapter closes with a discussion of some methodological issues of the study.

* i.e. the breeding and rearing of young pigs to 8 weeks; they then go on to the fattening stage which is not covered in this report.

2. The development of intensive livestock production

The past three decades have witnessed enormous changes in live-stock production. Rising labour costs have stimulated re-organisation to economise on labour and permit the mechanisation of many tasks. This has increased the number of animals which one stockman can manage and has reduced the gross margin per animal, providing an incentive to increase unit size. By 1978, 57 per cent of layers were housed on units of over 20,000 birds, compared to only 9 per cent in 1965.[3] The average number of sows on a full-time holding more than doubled between 1965 and 1975, from 34 to 80.[4]

A preoccupation with improving control over the environment and behaviour of the livestock, together with the need to accommodate larger concentrations of animals, led to a move away from outdoor systems. So the proportion of dry sows kept in paddocks for at least part of the year approximately halved from around 30 per cent to perhaps 15 per cent or less over the decade to 1980. Free-range laying birds predominated in 1956, accounting for 44 per cent of the total. By 1966 their share had fallen to 13 per cent, and by 1980 to less than 2 per cent.[5]

Once animals had been moved indoors, the need to economise on costly space and the opportunity to mechanise further promoted greater intensification. The deep-litter systems, which had replaced free-range for egg production, were overtaken by batteries in the mid-1960s which have gone on to house 96 per cent of all hens today. Similarly, the comparatively spacious grouped and straw based indoor pig systems are being replaced by individual or small group housing, more tightly confined and based on slats. Multiple-suckling— the grouping of sows and their litters in large straw yards— declined during the 1970s from a half to less than a third of all sows. There has been a corresponding rise in the proportion of tethered sows amongst MLC* recorded herds, increasing from 12 to 30 per cent between 1972 and 1980. Slatted floors amongst feeding herds rose frome 27 to 40 per cent over the same period.[6]

The introduction, post-war, of artificial fertilisers loosened the dependence of arable farmers on livestock waste. Similarly, the introduction of specially formulated and tested compound feedstuffs meant that livestock farmers turned to the feed merchant for their feed rather than growing their own. The new emphasis on the application of scientific knowledge to farming, and the increased capital-intensity of production, added to the pressure to specialise.

Alongside intensification, the livestock industry has also achieved rapid advances in production efficiency. In the EEC the average yield

* Meat and Livestock Commission.

per laying hen has risen from 181 eggs per annum in 1970 to 235 in 1977. In the UK the feed conversion rate for broilers was 3.2:1 in the early 1950s compared to 2.1:1 today.[7] In pig production, between the mid-1950s and the end of the 1970s, one annual survey noted an increase of almost a third in the number of pigs reared per sow, and feed-efficiency in both rearing and fattening has improved greatly as well.[8]

Real prices of livestock products have declined considerably over the post-war period. This is especially true of poultrymeat and eggs, which have also become cheaper relative to redmeat. It is not surprising therefore that consumption of pigmeat and poultry products have increased vastly during this century. Furthermore, despite the need to provide large quantities of grain as animal feed, increased levels of agricultural self-sufficiency have been achieved while population has increased by around a half.

We must, of course, avoid jumping to the conclusion that only intensification has caused improvements in productivity. As we show in the ensuing chapters, productivity has frequently increased in parallel under a range of conditions. This theme is returned to at the end of the book.

3. Some methodological considerations

There are many pitfalls to be aware of when making comparisons between the costs of different systems. Frequently, modern intensive techniques tend to be introduced by progressive, commercially acute farmers, using new buildings and with extensive advisory support. Less-intensive techniques are often practised by more traditional, non-specialist farmers. This poses particular difficulties when interpreting survey data. An extreme version of this dilemma is seen in the MAFF Egg Yield Survey which, because of the difficulty of finding sufficient numbers of commercial-scale, non-cage producers, tends to compare large commercial battery flocks with small 'backyard' non-cage producers. The latter may keep fewer than 100 birds, of indeterminate age, housed in makeshift unlit buildings. Generally stated, the problem is: to what extent does the comparative performance of systems reflect their inherent characteristics rather than other factors such as management ability or the quality of the buildings? Results of controlled experiments are useful in providing data for conditions in which management and other factors are, as far as possible, standardised. However, there is always a question mark over how far experimental conditions reflect commercial practice.

Incomplete knowledge of the technical and economic potential of different methods of husbandry results in further difficulties in making comparisons. Modern techniques may be in their infancy and, when fully developed and exploited, might show greater advantages than are now apparent. This may be true of early weaning, for example. Conversely, the focus of research, innovation and advisory effort over recent decades has been upon intensive techniques, which has often affected the rate at which less intensive techniques have been developed. It is very difficult to establish the net effect of these factors.

In comparing systems the present study endeavours to take account of difficulties such as these. Comparisons have been made in terms of what might be expected from new investments in a commercial enterprise, incorporating up-to-date equipment and techniques, and employing good stockmen and management. This will tend to reflect producers performing rather above average, and is fairly typical of many of the larger producers who dominate both egg production and breeding pigs.

A major part of the study is concerned with the development of 'theoretical' costings to allow comparison between systems; that is, estimates of the costs of production under different systems based upon imputed relationships between systems, feed efficiency, and the use of labour, capital and other resources. The 'theoretical' costings approach has been chosen because there are difficulties involved in obtaining farm survey data, and when it is available it can give only a historical perspective, from which generalisation needs to be undertaken cautiously. Theoretical costings can be designed for new investment in a range of systems, taking account of the latest techniques, and such factors as motivation and scale.

However, theoretical costings should not be taken too literally. They can only be as good as the data and judgments from which they are derived. They are inevitably stylised, and the comparison of average costs of production can conceal wide variations in practice. If there were to be a transition from contemporary intensive production methods to alternatives of the kind discussed here this would have important cost implications which are addressed in Chapter 5. Despite this, focussing upon the average production costs is a legitimate simplification because this is almost certainly the most useful indicator for comparisons over the fairly limited range of systems considered here. Our attention is given to commercially applicable systems and, as the following chapters show, the kind of changes with which we deal, while they may have important implications for animal welfare, do not imply radical changes in production costs or the pattern of resource use.

Of course, it cannot be assumed that differences in production costs will be automatically reflected in prices. The effect on prices of a widespread change in methods of production will depend not only on changes in average costs, but upon changes in the shape of industry cost curves, and the nature of demand. Nonetheless, from the point of view of the consumer, the farmer's production costs are of less interest than the retail price of eggs and pig meat. For this reason, the implications for retail prices of different production systems are drawn out in this report. However, because only the breeding stage of pig production is considered here, the discussion in Chapter 2 is confined to a comparison of production costs.

In deriving the costings presented in this report, use was made of a range of published surveys and experimental results. In addition, a number of farms were visited and two small surveys were undertaken to provide further information. The first survey was concerned with the performance of non-battery egg producers and the second with the price and retail margins prevailing in shops selling free-range eggs.

2. Breeding Pigs

1. The Pig Farming Cycle

Figure 1 introduces, in a simplified form, the pig farming cycle.

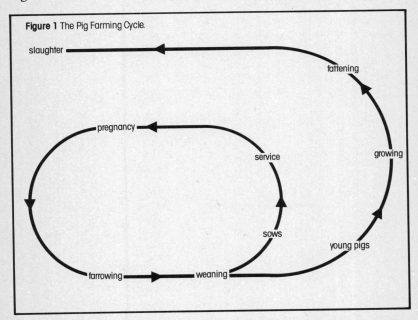

Figure 1 The Pig Farming Cycle.

slaughter

fattening

pregnancy

service

growing

sows

young pigs

farrowing

weaning

Sixteen weeks after service, the sow gives birth to a litter numbering perhaps 10 or 12 pigs, although this can vary greatly. Piglets are weaned at anything from 2-8 weeks; traditionally weaning is at 6-8 weeks, but there has been a strong trend towards earlier weaning during the past decade or so. This has been permitted by the development of high quality 'creep' feeds, providing an adequate nutritional substitute for milk, and greater control over the vulnerable young pigs' environment sufficient to replace maternal care. The sows generally come on heat three days after weaning and subsequently at 21 day intervals, and are served again, thus completing the cycle. Meanwhile, the 'weaners' enter a 'growing' stage. Some gilts, as the young sows are known before their first litter, are selected to replace or augment the breeding sows kept on the farm, while the remaining growing pigs go on to be fattened for slaughter.

Amongst large farm animals, breeding pigs have led the trend towards intensive husbandry. Today, perhaps only 5 per cent of sows are kept outdoors. Considerably fewer are outdoors for the whole of the breeding cycle, and for growing or fattening pigs to be outside is rare indeed! Indoors, the dominant trends have been to higher stocking densities, mechanisation of feeding and waste removal, greater environmental control through mechanical ventilation and some heating, more specialised buildings—often prefabricated or 'containerised', more concentrated feeds, earlier weaning and larger, more specialised units.

We have already discussed the trends underlying this intensification of production. We look now at the range of housing in which pigs are kept. Our attention is restricted to the breeding stage, which entails the production of young pigs up to the age of two to three months, after which they go on to the fattening stage.

2. Types of housing

2.1 Indoor housing

Pig buildings are designed to achieve a number of objectives. They must create an environment in which the pig performs efficiently; they must minimise the capital cost per pig; and they need to be economical in their use of labour. However, the ways in which intensive and semi-intensive designs seek to achieve these objectives are rather different.

Under intensive conditions, the temperature inside the building is kept high both by stocking the pigs densely so that they produce a high heat output, and by insulating the structure well in order to keep the heat in. Ventilation needs to be closely controlled and is usually fan-assisted so that it can be kept to a minimum in the winter to avoid heat loss but is sufficient in the summer to prevent over-heating. These requirements result in a high capital cost per square metre of floor area, re-inforcing the incentive to keep stocking densities high.

To achieve high densities, floors can be slatted and straw bedding dispensed with. Both imply a saving of space. However, pigs kept at high densities are inclined to be aggressive towards one another and it may be necessary to restrict their movement to prevent this. Dry sows, for example, are tethered or put in stalls. This adds further to capital cost, while restrictions on movement, together with the absence of straw bedding, make it essential to keep the building temperature high.

Control of the overall environment is of less importance in semi-intensive systems. Where straw bedding is provided and pigs are in groups they can keep warm by huddling together and burrowing into the straw, even when the overall building temperature is low. Furthermore, a kennelled lying area is often provided within the building; in this area temperatures are higher because of better insulation and a comparatively small air space and low level of ventilation. The pigs are then free to dung and exercise elsewhere in the building.

With less-intensive buildings the emphasis is on cheap forms of construction in order to compensate for the low overall stocking density. This can keep capital costs per pig at a competitive level.

A great many systems exist in practice, some of which combine 'intensive' and 'semi-intensive' elements. In Figures 2, 3 and 4 the main features of typical intensive and semi-intensive accommodation are illustrated for different stages of the pig cycle. There is further detail in Appendix 1.

2.2 Outdoor accommodation

In traditional systems, outdoor breeding herds are provided with fairly rudimentary accommodation. At its simplest, this can be uninsulated half-round corrugated iron huts, with a wooden or straw-bale barrier at the rear, and perhaps across part of the front also. Typically, dimensions are about six feet by nine feet, with a minimum height of three feet. The simplest huts have no floor, and are bedded with plenty of straw. It is important that the bedding is kept dry, requiring light, well-drained land. A farrowing sow requires a hut to herself, whereas dry sows can share one hut between six. A retaining board may be used to confine straw and very young pigs to the hut. This very simple accommodation is popular amongst farmers who have outdoor herds.

More sophisticated farrowing accommodation, known as the Craibstone Ark (Fig. 5), can also be used. Of double-skin wooden construction, with a floor and baffled entrance, this provides a more protected environment and makes observation of the sow and litter easier. A creep area and nest can be included, and supplementary heating can be provided, perhaps using portable bottled gas. The disadvantages are expense (although still far below that of indoor farrowing accommodation), and some loss of portability. For these reasons Craibstone Arks are less widely used than the simpler huts.

Recently, a large outdoor pig producer has been experimenting with mobile farrowing containers. These are well-insulated, comprising four

10

Figure 2 Dry Sow Accommodation

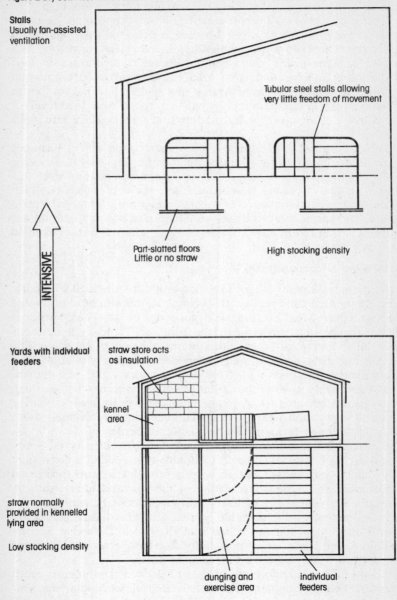

Stalls
Usually fan-assisted
ventilation

Tubular steel stalls allowing
very little freedom of movement

INTENSIVE

Part-slatted floors
Little or no straw

High stocking density

**Yards with individual
feeders**

straw store acts
as insulation

kennel
area

straw normally
provided in kennelled
lying area

Low stocking density

dunging and
exercise area

individual
feeders

Figure 3 Farrowing Accommodation

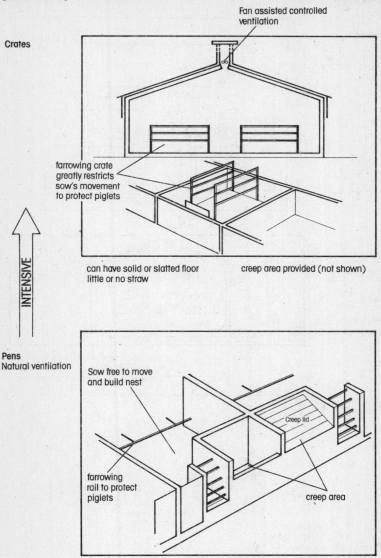

Crates

Fan assisted controlled
ventilation

farrowing crate
greatly restricts
sow's movement
to protect piglets

INTENSIVE

can have solid or slatted floor
little or no straw

creep area provided (not shown)

Pens
Natural ventilation

Sow free to move
and build nest

Creep lid

farrowing
rail to protect
piglets

creep area

straw bedding provided

12

Figure 4 Rearing Accommodation

Flat-decks

often heated for younger
piglets fan-assisted
ventilation

INTENSIVE

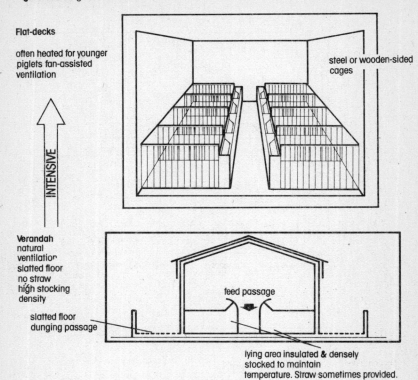

steel or wooden-sided
cages

Verandah
natural
ventilation
slatted floor
no straw
high stocking
density

slatted floor
dunging passage

feed passage

lying area insulated & densely
stocked to maintain
temperature. Straw sometimes provided.

Weaner pool
natural ventilation

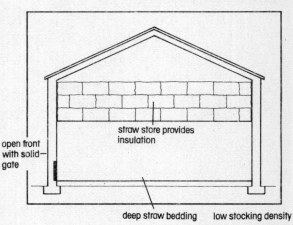

open front
with solid—
gate

straw store provides
insulation

deep straw bedding   low stocking density

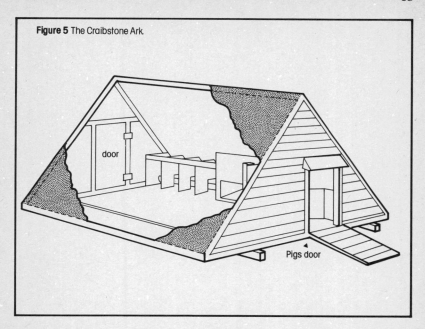

Figure 5 The Craibstone Ark.

door

Pigs door

pens fitted with farrowing crates and heated creep areas. Although expensive to buy, the cost might be justified if heating and straw costs are reduced and performance improved. The containers provide indoor-type accommodation which can be used on a mobile and wholly outdoor unit.

In most herds, sows are moved to fresh ground to farrow, and stocked at 4-6 per acre, although this will vary according to the quality of the land and the season. Dry sows are often stocked at a similar rate, although they may be kept at densities of 10-15 per acre. Fencing may be electric or post and wire-netting. Often the latter is used for perimeter barriers, with electric fencing sufficient for internal divisions, although many units now rely on electric fences alone.

Traditionally, outdoor systems have weaned at 5-8 weeks. However, recently they have begun to follow the general trend to earlier weaning and three week weaning has been successfully introduced by a number of farmers. Weaners are then moved indoors into either controlled environment accommodation, or kennelled, straw-based housing.

14

3. Production costs

In this section we examine the main influences upon feed, labour and
capital requirements and try to identify how they might vary between
systems. As explained earlier, the focus is on production, rather than
retail, costs.

3.1 Feed Costs

Feed accounts for well over half of the cost of producing breeding pigs.
Many factors influence feed costs, but the one most frequently cited in
comparing systems is the thermal environment. We consider this aspect
in detail here.

Put simply, 'cold' pigs eat more in order to maintain body temperature.
As a by-product of metabolic processes, pigs produce heat. Some of this
is required to maintain their body temperature. Additionally heat is lost
to the environment, at a rate depending largely upon how warm that
environment is, the air-speed, the thermal conductivity of surfaces with
which the pigs are in contact, and the size of the pig. Heat is lost more
rapidly in cold draughty conditions than in a warm and still environment.
To some extent, a pig can regulate the rate at which it loses heat. By
wallowing, for example, it can increase its heat loss through evaporation.
By huddling up to other pigs, it can reduce the surface area of its body
from which heat is lost. The rate of heat loss is also reduced by
burrowing into deep straw-litter.

Pigs kept singly and without straw are less able to make these
behavioural adjustments to regulate heat loss.

However, in spite of behavioural adjustments, there is a temperature
below which heat loss will exceed heat production, known as the *lower
critical temperature* (LCT). Below this temperature, an animal cannot
sustain bodily functions indefinitely without consuming feed directly to
produce heat, so as to maintain a balance between the heat which it is
producing and that which it is losing.

A pig's LCT will vary according to a number of factors, principally:
— The size of pig; smaller pigs are more vulnerable as they have a
higher ratio of surface area to body weight.
— The ventilation rate. Higher rates of air movement increase
convection losses;
— The plane of nutrition. Higher feed intake results in greater
heat production;
— Group size. Pigs in groups can huddle to reduce surface area
exposed to heat loss;
— The thermal conductivity of surfaces with which the pig is in

contact; well insulated surfaces reduce conductive heat loss. Unfortunately, LCT estimates are not very precise. This is not surprising in view of the number of specific environmental factors upon which they depend; but these are not always made explicit when estimates are presented. On the basis of available estimates, we present a range in Table 1.

Table 1. ESTIMATES OF LOWER CRITICAL TEMPERATURE FOR GROUPED GROWING PIGS AND SINGLE SOWS IN THE ABSENCE OF STRAW-BEDDING

Liveweight (kg)	LCT* Range (°C)	
growing pigs:		The lower estimate of LCT generally reflects
20	16-24	conditions of low air-speed. Variations in air-
40	10-21	speed are capable of accounting for the disparity
60	11-18	between estimates.[7] The figures for dry sows'
80 ⎫		LCT do not take account of the possible effect
100 ⎬	10-18	of group size. The extent of this effect will be in-
120 ⎭		fluenced by feed level and floor type. It has been
dry sows	10-18	claimed that a 140 kg sow individually housed
farrowing sows:	13-18	on straw bedding has a LCT about 4°C higher
		than a group of 5 sows kept under similar
		conditions.[8]

Sources:[1] − [6]

Straw and the thermal environment. There is considerable evidence which demonstrates that the provision of straw bedding has a substantial influence upon a pig's LCT. The type of floor on which pigs are kept influences their rate of heat loss. Straw-bedded pigs can be kept at lower temperatures than pigs without straw with no adverse impact on feed conversion. Stephens (1971) found that, compared with straw bedding, concrete increased heat loss from piglets by an amount equivalent to a reduction in environmental temperature of 10°C[9]. With fattening pigs of 40kg weight, Verstegen and van der Hel (1974) found that, compared with straw bedding, concrete slats had an effect on heat production similar to lowering the air temperature by 7.5°C[10]. Danish research work has shown that groups of finishing pigs kept at 3°C with a deep bed of straw converted feed as efficiently as groups without bedding at 8°C[11]. This suggests that good straw bedding is equivalent to 5°C, at low temperatures.

Bruce and Clark (1979) have developed a model of critical-temperature for growing pigs. They comment: 'The effect of straw is to reduce the critical temperature on concrete by about 6°C reducing to 5°C at the higher liveweights ... the difference between concrete and straw is more or less independent of liveweight and group size but increases with feed intake [12] .' Bruce (1979) has illustrated graphically the relationship between feed intake, floor-type and air-temperature for a pregnant sow, from which Table 2 is derived [13].

Table 2. COMPARABLE TEMPERATURES FOR A 170 KG SOW ON STRAW OR CONCRETE FLOORS AT VARIOUS LEVELS OF FEED INTAKE

°C		
Straw	Concrete	Feed intake (kg/day)
18	22	1.75
13	18	2.0
10	16	2.25
4	12	2.75

The advantage of straw can be equivalent to as much as 8°C, although it is likely to be in the region of 5-6°C under normal commercial conditions.

On the basis of all these results a conservative estimate is that the provision of straw corresponds to an increase in air temperature of 5°C for sows, and 6°C for growing pigs. Where comparatively low temperatures or young pigs are being considered, the effect of providing straw may be considerably greater.

Temperature and feed intake. Having established pigs' LCT under a variety of conditions, the amount of extra feed required by pigs kept below their LCT needs to be determined.

Verstegen and van der Hel (1974) estimate that the extra feed required is 0.3g per kg liveweight per °C below LCT. [14] . Estimates made by Holmes and Close (1977) suggest that the relationship varies according to bodyweight, as Table 3 shows.

Simplifying, (on the basis of Table 2) the following estimates of extra feed consumption per animal per degree-day* below LCT (based on Holmes and Close) can be adopted:

20 kg weaners	0.014 kg
40 kg store	0.022 kg
pregnant sows	0.053 kg

* a degree-day is the number of degrees below a base temperature multiplied by the number of days that this temperature prevailed. Thus, for a base temperature of 20°C, 2 days with an average temperature of 10°C would equal 20 degree-days.

Table 3. INCREASE IN FEED CONSUMPTION PER DEGREE-DAY* BELOW LCT

Liveweight (Kgs)	Extra feed required per degree-day (individual pigs)	
	grams per day	grams per kg liveweight
20	14	0.70
60	26	0.43
100	36	0.36
140 fat	34	0.24
140 thin	59	0.42

Source: derived from Holmes and Close (1977).[15]

Effect of housing on feed consumption. With this information, the effect of housing type on feed consumption can be depicted in very simplified form. Assume, for simplicity, that the temperature in a controlled-environment house is maintained at, or above, the LCT. Assume that the ambient temperature prevails in less intensive forms of housing, but that the pigs have the advantage of deep-straw bedding and, in the case of dry sows, being able to huddle together in groups. Some estimates of the effect of adopting these assumptions for a range of LCT values are shown in Table 4.

It must be emphasised that the assumptions upon which these results are based are very unfavourable towards the non-controlled environment buildings. In practice, most will maintain temperatures above those prevailing outdoors, particularly if they are insulated and provide some control over ventilation. Furthermore, it is optimistic to assume that controlled environment buildings do not allow temperatures to fall below the pigs' LCT. It is a common criticism of dry sow housing using tethers or stalls that it can be too cold and that design weaknesses can result in draughts and cold areas.[16] [17]. Flat-deck accommodation, too, does not always maintain optimal conditions in practice.[18]

Even with these unfavourable assumptions for less-intensive housing, the difference in feed consumption is not as great as is commonly supposed. The greatest disadvantage, predictably enough, comes in the feeding of young pigs or around 20kg, which might consume 5-10 per cent more feed. Store pigs and pregnant sows are likely to eat up to 5 per cent more than those kept under optimum conditions.

A revised model. However, it is clearly unsatisfactory to assume that non-controlled environment buildings are incapable of conserving heat. Many designs are well insulated and are provided with means for

Table 4. ESTIMATES OF EXTRA FEED CONSUMPTION ENTAILED IN LESS INTENSIVE BUILDINGS ACCORDING TO A SIMPLIFIED MODEL

Pig	LCT (°C)	less group/ straw effect (°C)	'average' UK climate % extra feed	adverse UK climate % extra feed
weaner — 20 kg	18	12	4.9	6.0
	20	14	6.6	7.9
	22	16	8.6	10.1
store — 40 kg	12	6	1.4	1.7
	15	9	2.7	3.4
	18	12	4.7	5.6
sow — 160 kg	12	3	1.4	1.8
	14	5	2.5	3.2
	16	7	4.2	5.4

influencing the ventilation rate. It is possible to make the model more realistic by taking some account of the internal environment of pig buildings, using established techniques. These enable the *difference,* between internal building and external temperatures, to be predicted from knowledge of the amount of heat given off by the housed pigs, the stocking density, the ventilation rate, and the building's insulating qualities (see Appendix 2). However, it is important to be aware that any such analysis is subject to a number of limitations. Although widely used, the model of the thermal properties of buildings is a crude one, reducing highly complex relationships to a small number of variables. Not much is known about the thermal behaviour of naturally ventilated farm buildings, which can be complicated by the existence of areas such as kennels which have different properties to the rest of the buildings. Ventilation rates are particularly difficult to estimate and yet are enormously important. The actual behaviour of similar buildings can vary widely according to where they are sited, and can also be influenced by alignment, degree of shelter and local climate.

This model can be used to refine our earlier results for feed consumption under less intensive conditions. It can be related to the thermal behaviour of a typical less-intensive type of pig house, a kennelled lying area with a separate exercise/dunging area. Pigs will spend most of their time in the lying area; it has been observed that fattening pigs spend 80 per cent of their time lying [19] , and there is no obvious reason why the behaviour of sows should be greatly different. We can assume that the yarded area is at the ambient temperature, although it may not actually be outdoors. The kennelled area can be well insulated, and might have a wall extending across the front to cut down on excess ventilation (it is assumed here that the front is three-quarters walled).

Table 5. DIFFERENCE BETWEEN INTERNAL AND EXTERNAL TEMPERA-TURES FOR A KENNEL-TYPE BUILDING AT DIFFERENT VENTILATION RATES. (°C)

Pig	min. recommended ventilation rate	2 x min. recommended ventilation rate	3 x min. recommended ventilation rate
weaner —			
20 kg	16.5	9.1	6.2
store —			
40 kg	12.6	6.9	4.7
dry sows	9.5	5.1	3.5

The results are presented graphically in Figure 6. They show that even with very unfavourable assumptions about ventilation rates, only small amounts of extra feed are required in kennel-type buildings compared to optimal conditions.

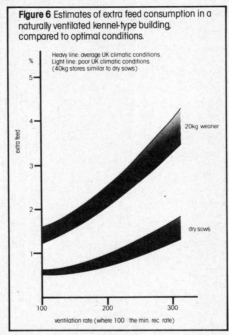

Figure 6 Estimates of extra feed consumption in a naturally ventilated kennel-type building, compared to optimal conditions.

Heavy line: average UK climatic conditions.
Light line: poor UK climatic conditions.
(40kg stores similar to dry sows)

Some empirical data. The Pig Demonstration Unit of the National Agricultural Centre, (NAC) records for comparative purposes the results obtained under a variety of systems operated there. Although it is a demonstration centre, it is also a commercial unit and has the advantage of operating different systems under the same management; results collected from many different farms can confuse intrinsic differences in systems with variation in the quality of management. The NAC have recently begun comparing flat-decks with first-stage kennels, for pigs of 5-16 kg. The results for 1979, the first full year for which results are available, are presented in Table 6.

Their Report comments: 'A comparison between the 1979 results for the flat-deck and kennel systems shows a remarkable similarity in terms of physical performance. It demonstrates that two completely different systems of 3 week weaning can work equally well given good levels of

Table 6. FOOD CONVERSION RATE & MORTALITY COMPARED FOR FLAT-DECKS & FIRST-STAGE KENNELS (1979)

| | Flat-decks | First-stage kennels | |
		Slatted runs	solid runs
FCR	1.5	1.5	1.5
mortality (%)	1.1	1.6	0.6

stockmanship and an adequate thermal environment for the pigs'.[20].

In an ADAS study of piglet performance between three and nine weeks of age in weaner houses, carried out between October 1975 and August 1977, 42 groups of pigs were monitored on 20 farms around the country, in various types of buildings, classified according to type of floor and system of ventilation. No firm conclusion can be drawn because of the diversity of conditions and small size of the sample— for example only four naturally ventilated buildings were studied. However, from the results there appeared to be little correlation between piglet performance and any of the environmental factors, except in the case of daily gain and floor area per pig, which were positively correlated. Satisfactory results were obtained in various types of buildings and in some cases it appeared that good stockmanship had, almost predictably, overcome deficiencies in the housing.[21] Three of the naturally ventilated buildings showed good results, the other suffering an outbreak of scour during the recording period. Commenting on the results obtained in the naturally-ventilated buildings during the winter of 1976-77, the Report states: 'Both units used straw and the results confirm that good performance can be achieved even when there is a wide range of temperature, provided that management is good and the pigs can sleep in a favourable micro-climate.'

In conclusion, although-thermal environment does affect feed consumption, in practice the thermal environment in a less-intensive house is unlikely to be significantly poorer than in a controlled-environment building. Where a good straw-bed is provided this is equivalent to an increase in air temperature of perhaps 5°C for sows, and more for growing pigs. Grouped pigs will huddle together when cold in order to reduce their heat loss, behaviour denied to pigs in stalls. And less-intensive housing can incorporate a well-insulated kennel as a lying area, with a small air-space and a low ventilation rate in order to conserve heat. As a result, differences in feed consumption due to temperature are unlikely to be large between well managed systems.

3.2 Capital Costs

Capital costs are likely to account for between 5 and 15 per cent of breeding costs.

In making estimates of capital costs it is important to be aware that they are likely in practice to differ greatly from farm to farm, depending upon such factors as distance from the supplier, characteristics of the site, the extent to which farm labour can be used, layout and standard of construction. Even where overall building costs are similar, differences in stocking density— particularly where the systems are strawbased— can mean substantial variations in cost per pig. Drawing upon information from several sources, estimates of capital costs are presented in Table 7.

Capital costs may be divided into two: the depreciation in the value of the assets; and the interest charges incurred on capital tied up in the investment. Even where the farmer provides the capital so that no interest is paid, the capital could have earned a return elsewhere, and this opportunity cost is therefore included in comparisons in the form of an imputed interest payment.

For the present analysis an annual capital cost has been calculated, based on a given rate of interest and an assumed depreciation period.

It has been assumed that investment should provide a real* rate of return of 5 per cent. This is consistent with, and probably a little higher than, existing returns on tenants' capital [22] and is in line with real interest rates prevailing over the past two decades.

The speed with which capital depreciates depends both upon physical ageing and obsolescence. Specialist pig buildings may become obsolete if they are not adaptable to new techniques, or if they cannot be used efficiently for other livestock in the event of a farmer leaving pig production. In calculating annual costs, a depreciation period of 15 years has been assumed for most buildings. However, 10 years is assumed for the building and equipment required for outdoor production to allow for greater wear and tear where huts are regularly moved and a high proportion of items, such as fencing, which depreciate comparatively rapidly. Accommodation which makes use of general purpose structure (sow yards, multi-suckling and weaner pools) is depreciated over 20 years to reflect its greater adaptability to other uses. The capital costs set out in Table 7 only indicate the cost of the individual components of a system. It is not necessarily obvious from this how costs compare, as the way in which buildings are utilised varies. For example, the cost of a solari farrowing unit is less than a crate house. However, if pigs in a crate house were to be weaned earlier than in the solari unit, more farrowings could be obtained in the crate house. This would affect the relative cost

* i.e. adjusted to exclude the effects of inflation

Table 7. INITIAL CAPITAL COSTS OF PIG PRODUCTION SYSTEMS—AUTUMN 1980

Dry Sows	£ per pig place
Stalls—part slatted	260
Tethers—solid floors	210
Cubicles	310
Kennel & yard—individual feeders	330
Deep-straw yard—individual feeders	380
Outdoors—½ round huts incl. fencing drinkers, etc.	25
Farrowing & Lactation	
Cratehouse—part slatted	700
solid floor	650
Solari-type	600
Group-suckling	400
Outdoors—½ round uninsulated huts	140
—insulated with creep heater	260
—container	600
Rearing	
Multi-tier cages	60
Flat decks	45
Weaner verandahs	28
Weaner pool—deep straw	50
Monopitch	24

Sources to Table[23]–[29]

per farrowing, depending on the expense of the post-weaning accommodation for the sow and litter. Thus it is necessary to define all of the accommodation required for specific breeding systems before true capital costs can be compared.

The number of possible combinations of accommodation is large, so attention will be restricted to a few illustrations. In Figure 7 the effect of weaning age on accommodation costs is calculated and Figure 8 shows the effect of different dry sow accommodation on overall capital costs.

As Figure 7 shows, there is a remarkable similarity between the total capital costs per sow at different weaning ages. The lower cost of farrowing accommodation is offset entirely by higher sow and weaner costs for earlier weaning systems. Weaning age barely affects capital cost per sow.

These capital costs do not include provision for ancillary items such as boar accommodation, feed storage, isolation and office buildings, roads and passageways, fencing, mains service and waste storage. These

24

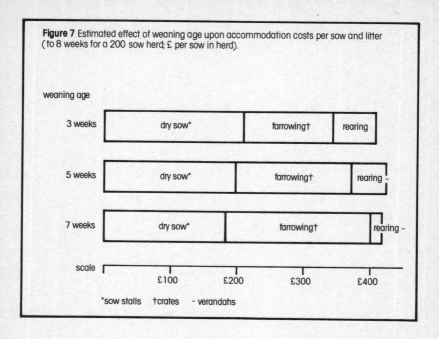

Figure 7 Estimated effect of weaning age upon accommodation costs per sow and litter (to 8 weeks for a 200 sow herd; £ per sow in herd).

*sow stalls †crates ‑ verandahs

costs can vary enormously and are likely to add between 50 and 100 per cent to the capital required to set up an indoor unit. Because of the wide variation, estimates have not been developed here.

Figure 8 indicates that, indoors, dry sow housing accounts for a large part of housing costs. The disparity in dry sow housing is quite large, with stalls costing £210-260 and yards about £330-£380 per sow place. Figure 8 also illustrates that quite large savings in initial capital expenditure can be made by choosing either stalls or outdoor housing. However, looking at annual costs, the disparities are much less significant. Only outdoor housing appears to offer any great saving in annual capital costs, even when rent is allowed for.

3.3 Labour use

Labour costs are not a large proportion of total costs, accounting for between 10 and 15 per cent in breeding herds. Nonetheless, considerable attention has been directed at reducing labour requirements. Real wages have risen rapidly in relation to other costs and the value of output, providing a strong incentive to economise; the value of a bacon

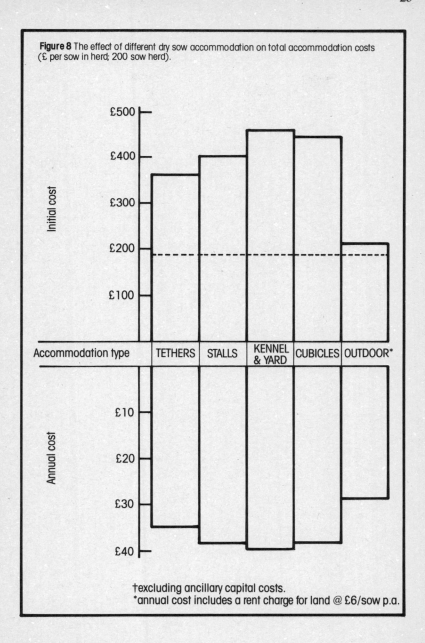

Figure 8 The effect of different dry sow accommodation on total accommodation costs (£ per sow in herd; 200 sow herd).

†excluding ancillary capital costs.
*annual cost includes a rent charge for land @ £6/sow p.a.

pig was equivalent to the cost of about 142 hours of work in 1953-54, but only 34 hours in 1978/79[31].

Labour use can vary widely between farms using very similar systems, due to differences in design, management, ability and motivation. Indeed, farms using similar management systems often vary as much as those using different systems.

However, labour requirements can be an important element in the choice between systems, especially where margins are low, and intensification has been correlated with a reduction in labour required per pig. To some extent this has been due to modernisation and specialisation rather than the use of more intensive techniques. Thus specialised pig buildings designed in an era of high wages will require less labour than buildings converted for pig production on an ad hoc basis to utilise surplus farm labour, even if the husbandry methods are identical. Complications of this kind would make precise identification of the relationship between labour costs and intensification acutely difficult even if detailed data on labour costs existed— which it does not.

Furthermore, the relationship is not a simple linear one. A number of extensive systems can be very economical in their use of labour, especially where feeding is carried out in large groups and cleaning is either infrequent, where a deep straw litter is provided, or is eliminated, as is the case outdoors.

Feeding and cleaning are the major tasks on a pig farm and together can account for three-quarters or more of total labour needs on many holdings. It is here, therefore, that attention should be focussed.

Looking first at cleaning and bedding, Table 8, outlines some of the tasks involved with various systems of waste handling.

Time spent feeding is dependent upon many factors. Some of the more important are: the frequency with which pigs are fed; the number fed in a group; whether feeding requires entering a pen or can be carried out from a passage; the distance between feeding points; and the extent of mechanisation and automation. It is probable that well-designed systems can be economical in their use of labour, irrespective of how intensive they are. The process of feeding yarded sows in individual feeders is very little different from feeding sows which are in stalls, apart from securing and releasing sows from the feeders before and after feeding.

Provided that troughs can be reached from outside the pen, feeding sows and litters in a row of farrowing pens, such as the solari-type, should not take very much longer than walking from pen to pen inside a farrowing house. There is unlikely to be much difference in time spent feeding in different rearing systems. Feeding is generally ad lib and

Table 8. TASKS INVOLVED IN CLEANING AND BEDDING

Type of pen	Pen cleaning	Storage and disposal
Solid floor pen without dunging area	*either:* regular (usually daily) removal of dung from each pen, manually with barrow/trailer *or:* regular addition of fresh straw to form deep-litter, removed infrequently and where possible by tractor and fore-end loader.	removed to midden or compost bin for storage, from where it is returned to the land. Can be loaded directly from buildings into muck spreader when conditions favourable, reducing handling.
Pen with solid floored dunging passage	regular (usually daily) removal of dung by scraping with squeegee — manual or (more usually) tractor mounted. Common dunging passage allows many pens to be scraped in succession. Pigs need to be excluded from passage by closing gates while scraping takes place. Occassional cleaning of remainder of pen where pigs haven't used passage	generally scraped into slurry-lagoon, from which it is transferred to land when conditions permit.
Pen with partly/fully slatted floor	occasional cleaning of dung from slatted and solid areas. Most dung falls through slats into slurry channels.	stored in slurry channels or transferred — often by tanker (or pumped) to lagoon/ tanks outdoors. Thence by tanker (or sometimes irrigation pipes) to land as conditions permit.

group size quite large, which means that labour used in feeding is fairly minimal anyway.

There is remarkably little evidence available on labour use under different systems. Much of it is dated, and allows only limited and specific comparisons, from which it is difficult to generalize. Some observations of labour-use in farrowing accommodation are made below.

In their Bulletin, *Housing the Pig,* the Ministry of Agriculuture comments that for farrowing sows there is "surprisingly little difference between the time required for the various indoor systems". The first 3 weeks in farrowing accommodation take around 3-3½ hours of labour per sow, with crate systems closer to 3 hours and farrowing pens nearer 3½. Outdoors, a little more time is needed, estimated at about 3¾ hours over a 21 day period[32] It has been reported on the basis of work at Purdue University in the USA that slatted floors can reduce the labour requirements in farrowing pens by 60 minutes over a 21 day farrowing period; also, on the basis of comparative trials, that it takes 3.8 minutes less per day (80 minutes over 21 days) to feed and clean in a farrowing crate with a totally slatted floor than a similar pen with a solid floor[33]

However, a point to bear in mind is that in some cases examples of labour savings may, on closer examination, be illusory or exaggerated. Under all systems, regular inspection of pigs is very important. Indeed, where pigs' movements are particularly constrained by the housing, very diligent observation of tethers and slats, for example, is strongly recommended. Where the husbandry system requires frequent work with the pigs—bedding them down, removing and manual feeding—inspection is a part of these activities. Where these processes are dispensed with, inspection becomes an additional task, and potentially quite a time absorbing one.

Some idea of the range of labour costs in practice can be obtained from the Cambridge Pig Management Survey.

Labour requirements per *sow* are similar for all weaning ages and differ surprisingly little between the 'best' and 'worst' performing herds. Indeed variation between herds in the same group is large in relation both to total labour requirements and variations between groups, as the large standard deviations indicate. It is noticeable that small herds of less than 50 sows have markedly greater labour requirements, which may reflect upon results by weaning age, as later weaning herds tend to be quite a lot smaller than those weaning earlier.

The range in labour costs per *weaner* is rather greater because on average the groups of herds with the lowest labour costs also have comparatively high numbers reared per sow.

Table 9. LABOUR COST PER SOW P.A. IN CAMBRIDGE SURVEY HERDS (1980).[34]

	(standard deviations in parentheses)			
by performance*	Average	Best 20	Worst 20	
Labour cost/sow p.a. (£)	61.05 (17.95)	49.53 (14.63)	60.37 (16.38)	
per weaner (£)	3.23 (0.95)	2.37 (0.70)	3.87 (1.05)	
by weaning age	3 weeks	3 & 4 weeks	5 & 6 weeks	7 & 8 weeks
Labour cost/sow p.a. (£)	58.14	61.11	61.78	64.72
per weaner (£)	2.85	3.15	3.58	4.02
by herd size	50 sows	50-99 sows	100-199 sows	200 & over
Labour cost/sow p.a. (£)	76.99	61.86	60.80	60.12
per weaner	4.01	3.14	3.20	3.25

*selected on total cost per kg of weaner

In conclusion, although the more intensive systems tend to have lower labour costs, there are some extensive systems, especially outdoors, which can be very economical in their use of labour. Differences between systems are not large, often no greater than disparities between farms using similar systems, and because labour costs amount to less than 15 per cent of total costs in breeding herds they are not a particularly important influence upon overall cost differences between systems.

4 Management Practices and Performance

In this section, some of the management practices which have been brought up in the debate on animal welfare are examined, and their implications for the efficiency of pig rearing are assessed. The following issues are discussed: the relationship between weaning age and production cost; the value of farrowing crates; the comparative performance of group and single suckling; and the performance of herds breeding outdoors.

4.1 The effect of weaning age on production cost

In recent years, earlier weaning ages have become increasingly popular amongst pig-farmers, as table 10 shows.

Table 10. DISTRIBUTION OF AGES AT WEANING IN CAMBRIDGE SURVEYS

| | Age at weaning — weeks | | |
	3-4	5-6	7-8
			% of herds
1971	10	59	31
1975	18	67	15
1979	57 (i)	34	9

(i) Including 11 per cent of herds weaning at less than 3 weeks
Source: Pig Management Scheme Results, R.F. Ridgeon, University of Cambridge

There is considerable pressure to further reduce weaning age, even to the extent of experimental work on weaning at birth, although it is uncertain whether this practice would ever be considered suitable for commercial use. However, from the welfare viewpoint, it has been objected that early weaning deprives piglets of maternal contact and is stressful to the sow.

The principal advantage of earlier weaning is that it permits sows to have a larger number of litters in a year — which should increase the total number of pigs reared per sow. The *theoretical* possibilities are illustrated in Table II.

Table 11. THEORETICAL EFFECTS OF WEANING AGE ON PIGS PER SOW PER YEAR

Age at weaning (days)	21	35	56
weaning to service interval (days)	8	8	8
pregnancy (days)	115	115	115
TOTAL DAYS	144	158	179
Litters per sow per year	2.53	2.31	2.04
Pigs reared per litter	9.0	9.0	9.0
Pigs reared per sow per year	22.7	20.8	18.4

Source: MLC (1980) [35]

In theory, weaning at 3 weeks rather than 8 weeks can result in more than 4 extra pigs being reared per sow in a year. If this were achieved it could substantially reduce costs *per piglet reared*, by spreading labour capital and the cost of feed required for sow maintenance and replacement, over a larger number of piglets.

In practice, the picture is more complex. For one thing, performance does not attain the levels theoretically possible. In addition there are further factors to offset the savings made by earlier weaning. Sow performance may be worse, with fewer pigs reared per litter and at smaller weights, higher sow mortality, a greater weaning to service interval, and a higher number of returns to service.

Table 12. BREEDING RESULTS RELATED TO WEANING AGE: SURVEY RESULTS (1979)

Age at weaning (days)	Less than 18		19-32		33-46		47-60	
	Cam-bridge	MLC	Cam-bridge	MLC	Cam-bridge	MLC	Cam-bridge	MLC
Litters/sow/year	2.23	2.3	2.19	2.2	1.99	2.0	1.78	1.9
Pigs reared/litter	8.7	8.9	8.8	8.9	8.9	9.1	8.9	9.0
Pigs reared/sow/year	19.3	20.5	19.3	19.7	17.7	18.1	15.9	17.1
Extra pigs/sow/p.a. compared to weaning at 47-60 days	3.4	3.4	3.4	2.6	1.8	1.0	—	—

Sources:[36] [37]

Table 12 shows the breeding results obtained by farmers participating in two recording schemes. Weaning at under 4 weeks is associated with an extra 2½-3½ pigs reared compared to 7-8 week weaning, and an extra 1½ more than with weaning at 5-6 weeks. This is significantly less than is theoretically possible.

However, it is unlikely that these results show the *actual* relationship between weaning age and sow productivity. They will reflect other factors too. On the one hand, because early-weaning is fairly new to many farmers, their results are depressed by inexperience. On the other hand management and stockmanship are critical, and it is quite possible that standards vary according to weaning age. It is possible that it is the progressive, innovative farmers who adopt earlier weaning, and one might therefore expect the results of early weaners to be correspondingly better. In addition, as early-weaning is a relatively new commercial technique, it tends to take place on more modern units with, for example, well-organised service areas. Both these factors could contribute to the comparatively high performance levels. At the other end of the scale, later weaning farmers might be less able or willing to innovate, or simply anxious not to continually push their sows to the limits of their performance. If this is the case, then actual results by weaning age would overstate the intrinsic advantage of earlier weaning. It is extremely difficult to determine the net effect of these various and sometimes opposing factors.

Experimental data, where management factors are generally standardised, provide a way of overcoming some of these problems. The difficulty is, of course, that it is not always clear how far they can be taken to reflect commercial conditions. In a review of a large quantity of experimental and survey data from several countries, Brake (1978) attempted to define a relationship between reproductive performance, feed consumption, and weaning age. Taking into account differences in farrowing index*, numbers of piglets born alive, mortality, and proportion of gilts in the herd, the estimated number of 20kg pigs reared is shown in table 13[13].

Table 13. NUMBER OF PIGLETS (20 KG) PRODUCED PER LITTER & PER SOW AT VARIOUS LENGTHS OF LACTATION

Length of lactation (days)	Number of piglets (20 kg) per litter	Number of piglets per sow per annum
6-10	7.69	17.61
11-15	7.94	17.95
16-20	8.12	18.11
21-25	8.25	18.18
26-30	8.36	17.94
31-35	8.45	17.62
36-40	8.53	17.30
41-45	8.60	16.97
46-50	8.66	16.65
51-55	8.72	16.33
56	8.75	16.14

Source: (Brake 1978)

This suggests that weaning at 3-4 weeks leads to perhaps 2 pigs more reared per sow than weaning at 7-8 weeks, and around 1 pig more than 5-6 week weaning. (It should be noted that, due to rather smaller litters, the number of pigs reared is a little below results obtained in the UK.)

In the light of all these results a relationship between weaning age and breeding performance in the UK is proposed in Table 14.

The relationship can be summarised approximately in the following way:

For every week earlier that weaning takes place an extra half pig can be reared per sow p.a., over the range 3-8 weeks.

Breeding performance is only one half of the picture, and needs to be

*A farrowing index is the number of litters which a sow has in one year

Table 14. BREEDING PERFORMANCE RELATED TO WEANING AGE

Weaning age (weeks)	Litters per sow	Nos. reared/litter	pigs reared/sow/ p.a.
Top third herds			
3	2.35	9.2	21.6
5	2.17	9.4	20.4
7	2.02	9.6	19.4
Average herds			
3	2.20	8.8	19.4
5	2.05	9.0	18.4
7	1.90	9.2	17.5

viewed in conjunction with the costs involved in earlier weaning. The savings in sow feed are offset by the higher cost of the more concentrated creep feeds fed to piglets in place of suckling—creep feeds are roughly 50-100% more expensive than sow feeds; and earlier weaning of pigs requires more capital and energy intensive rearing accommodation, such as flat-decks and cages, to provide a closely controlled environment for vulnerable young pigs. Maternal regulation of the young pigs' environment is replaced by careful mechanical control.

Brake included feed useage and cost in his estimates, and they are presented graphically in Fig. 9. The number of pigs reared is greatest, and feed per piglet least, at weaning ages around 3 weeks. Feed *cost* per pig reared is least at around 4 week weaning, however, and the difference in feed cost is much less marked than differences in physical performance. Unfortunately, neither Brake nor the MLC survey consider costs other than feed costs, but data on all costs are available from the Cambridge Pig Management Survey, and are presented in Table 15.

Table 15. AVERAGE COST PER WEANER BY WEANING AGE (AVE. 1978-80): CAMBRIDGE PIG MANAGEMENT SURVEY[39]

Weaning age (wks)	<3	3 & 4	5 & 6	7 & 8
£				
feed	10.39	10.34	10.39	11.18
labour	2.57	2.75	3.00	3.51
other costs	3.49	3.22	2.88	3.08
stock depreciation	0.59	0.47	0.34	0.36
Total	17.04	16.78	16.61	18.13

34

The cost per weaner is very similar for weaning at 3-6 weeks, marginally greater for very early weaning, and substantially more for 7-8 week weaning. Stock depreciation is more rapid for earlier weaning, presumably due to a higher cull rate. Labour costs increase with weaning age, but it is highly likely that this is largely due to the use of more modern labour-saving equipment and buildings on earlier weaning farms. There may also be a size effect as later weaning herds tend to be substantially smaller than those weaning earlier.

On the basis of this discussion estimates of the relationship between

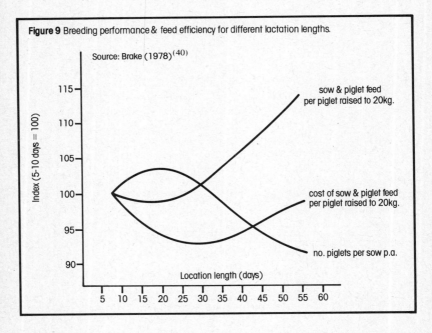

Figure 9 Breeding performance & feed efficiency for different lactation lengths.

Source: Brake (1978)[40]

weaning age and feed costs are presented in Table 16. This is inevitably speculative because of the difficulty of interpreting published data. It does, however, endeavour to reconcile the various sources to provide a plausible guide. The increased sow feed per weaner at later weaning ages is largely offset by reduced use of pig feed so that total feed costs at 3 and 5 weeks are very similar, while 7 week weaning increases feed cost per pig by about 7 per cent. There are, of course, costs other than feed which need to be included in an overall comparison. This is done later in Table 25.

Table 16. BREEDING AGE AND FEED COSTS BY AGE AT WEANING – SOME HYPOTHETICAL FIGURES

	Average herds			Top-third herds		
Weaning Age (weeks)	3	5	7	3	5	7
Farrowing index	2.20	2.05	1.9	2.35	2.17	2.02
Nos reared/litter	8.8	9.0	9.2	9.2	9.4	9.6
Nos. reared/sow p.a.	19.4	18.4	17.5	21.6	20.4	19.4
Sow feed/sow p.a. (t)	1.15	1.20	1.30	1.10	1.14	1.24
Feed/weaner to 8 Weeks (kg):						
Sow feed	59	65	74	51	56	64
Pig feed	24	21	18	23	20	17.5
Total	83	86	92	74	76	81.5
Ave. cost of feed/t (£)	159	155	153	155	152	149
Feed cost/weaner at 8 weeks:	13.18	13.41	14.05	11.49	11.56	12.14

Sources: MLC[41][42]; Cambridge Pig Management Survey[43]; Brake (1978).[44]

4.2 Farrowing: how necessary are crates?

Farrowing crates are made of tubular steel and limit sow movement almost entirely, allowing them only to sit, lie, or stand. In some cases even these movements are severely restricted. Larger sows' movements may be particularly inhibited, certainly where crate dimensions are not adjustable. Sows are put in crates up to a week before farrowing, and remain there generally until between 3 days and 3 weeks after farrowing.

The alternative to the crate is the farrowing pen, in which straw bedding is provided and there is more freedom for sows to nest and to move about. This system is no longer widely used.

While crates may be expected to allow some savings in space and labour, the overwhelming justification cited for their use is that they reduce piglet mortality—particularly by preventing sows from lying or stepping on their young. They are recommended as the most effective way of preventing such losses, and for this reason their use is accepted by some animal welfare groups. The overwhelming majority of farmers farrowing indoors have adopted crates; in the NFU Survey, 85 per cent of farmers used them either exclusively (71 per cent) or in conjunction with other systems (14 per cent).

However there are disadvantages too. Despite the widespread use of crates, piglet mortality remains substantial, approaching 15 per cent of pigs born alive[45][46]. And, according to an ADAS investigation of more than 2,000 farrowings in Lancashire during 1974-6, a third of pre-weaning mortality was still accounted for by overlaying[47]. The effect of restraint on behaviour during farrowing is being investigated by the Scottish Farm Buildings Investigation Unit. They have found that while unrestrained sows exhibit nest building activity, those restricted in crates carry out various stereotyped activities, interpreted as displaced nesting behaviour, accompanied by clear signs of stress. On the basis of work on other animals, demonstrating that stress delays and prolongs labour, typically leading to an increase in stillbirths, they suggest that this might also be true of pigs[48].

Crates clearly inhibit maternal care, transferring this responsibility to the stockman and his equipment. What needs to be established is the extent to which crates achieve an improvement in the number of pigs reared by each sow. Considering the importance attached to the use of farrowing crates, there is surprisingly little evidence to support this practice.

The evidence. One of the few pieces of experimental work on this subject was undertaken by Robertson et al (1966), who compared performance in a farrowing crate with that in a conventional pen which was equipped with a farrowing rail for piglet protection[49]. Their results are presented in Table 17.

Table 17. COMPARISON OF CRATE AND CONVENTIONAL PEN SYSTEMS

	Crate	Conventional Pen	Difference	Significance of treatment difference
No. of pigs— per litter:				
at 3 wks:	9.24	8.61	0.63	$P < 0.05$
at 8 wks:	8.90	8.03	0.87	$P < 0.01$
Mortality (%):				
to 3 wks:	15.54	21.27	−5.73	$P < 0.05$
to 8 wks:	18.65	26.57	−7.92	$P < 0.01$

(results presented as adjusted for variation in size at birth by Robertson, 1977[50])

There are some difficulties in interpreting these results. In the first place, they may be confused by factors such as temperature and management, which are not essential differences between the systems but which did differ within this experiment.

Secondly, no effort is made to consider the influence of selecting stock suited to each system, which one would expect to occur in practice. Large white sows and gilts were used, a breed not always noted for its mothering qualities, although it is possible to select within breeds for strains with these qualities. Thirdly, it appears surprising that the disparity in mortality is most pronounced, and statistically more significant, for the period beyond 3 weeks. This is not the period with which mortality from over-laying is most associated. In practice, farrowing crates are often not used beyond 3 weeks, particularly with the spread of earlier weaning.

With the above reservations, the advantage of using crates up to 3 weeks of age amounts to about two thirds of a pig in each litter, which is less than is often supposed.

A second source of evidence was provided by a survey of pig herds in Indiana, in which researchers attempted to establish the causes of piglet loss[51]. Regression analysis of their results indicated that, compared to pens, crates saved 0.41 autumn born pigs per litter and 0.22 spring born pigs. There is the possibility that subsequent improvements in crate design could have widened this gap.

In an unpublished analysis of mortality in the outdoor pig herds of

Baxter-Parker Ltd., Seidel (1980) compared their older farrowing huts, which provide heated creeps but not crates, with their newly introduced portable farrowing containers[52]. The containers have accommodation for 4 sows and litters each in crates and with heated creeps and are really mini indoor farrowing houses. A sow and litter remains in its crate for 1 to 2 weeks according to the season. Results for the period May 1979 to June 1980 are shown in Table 18.

Table 18. COMPARISON OF TWO KINDS OF FARROWING HUT, ONE WITH AND ONE WITHOUT CRATES

	New huts (with crates)	Old huts (without crates)	
Total farrowings	139	220	
% of these gilts	43%	18%	
Average born dead per litter	0.73	0.62	
Average born alive per litter	10.13	10.79	
Mortality per litter	1.19	1.37	
Mortality as % of pigs born	11.72	12.73	
Mortality due to:-	No. per litter	number per litter	number new < old (per litter)
crushing	0.48	0.85	0.37
scours	0.04	0.16	0.12
runts	0.35	0.25	−0.10
starved	0.09	0.03	−0.06
savaged	0.13	0.05	−0.08
Anaemia	0.08	nil	−0.08
other	0.02	0.03	−0.01
Total			0.18

The crude difference in mortality between the two systems amounts to only 0.18 pigs per litter. There is, however, a problem in comparing results as the new huts contain a higher proportion of gilts. It is possible that it is this, rather than differences between crates and huts per se, which accounts for the higher number of runts, starved and savaged pigs in the new huts. In this case the real difference in mortality between the two systems may lie somewhere between 0.18 and 0.36 pigs per litter.

Parrish (1972) carried out a survey of 25 pig herds, each with more than 500 sows in a single breeding unit. Of these, 18 farrowed indoors, and Parrish remarked: "If one subdivides the sows indoors into crate or pen farrowing then the result is an advantage for the crate by 0.54 extra pigs saved per litter[53]". Unfortunately he does not give details of the quality of the statistical information upon which this figure is based which makes it difficult to judge in isolation, but it is consistent with the other results which we have reported.

Almost all farmers farrowing out-of-doors do not use crates. The vast majority of indoor farrowings take place in crates. Therefore comparisons of indoor and outdoor farrowing results, whilst reflecting a number of factors, can cast some light on the advantages of crates. A useful source is Boddington (1971) who reported the percentage mortality amongst 41 outdoor pig herds during 1968/9. [54] At 12.82% of pigs born alive, this is equivalent to or, in most cases, *less* than the mortality amongst any of the four samples which he chose to represent the results of indoor herds. There is, however, a problem with the data. As outdoor farrowings tend not to be closely supervised, pigs dying between farrowing and first inspection are sometimes classified as born dead, and therefore wrongly excluded from piglet mortality. Unfortunately, Boddington did not collect data on numbers born dead, but he notes a tendency for substantially lower numbers of live births per litter amongst outdoor sows, which supports this interpretation. Yet, even making the assumption that equal numbers were in fact born alive, so that the smaller number of pigs apparently born outdoors was solely a reflection of recording errors, this would mean piglet mortality outdoors would still only be an extra 0.2-0.5 pigs per litter. From MLC surveys carried out recently it is not possible to discern any difference in mortality, or numbers born alive, between indoor and outdoor herds.[55]

The results from these various sources are drawn together in Table 19.

Table 19. NUMBER OF PIG DEATHS SAVED PER LITTER BY USING FARROWING CRATES

Source for estimate (see text for details)	
Bauman et al (1966)	0.22 - 0.41
Robertson et al (1966)	0.63
Seidel (1980)	0.18 - 0.36
Parrish (1972)	0.54
Boddington (1971)	0 - 0.5
MLC	0

With good management there is likely to be little advantage from using farrowing crates. These results show that crates may save about half a pig in each litter, although it is quite possible that the saving is substantially less than this. However, it must be remembered that crates can have other advantages for farmers; they make management more convenient and they are essential in a slurry-based system.

4.3 Group suckling

The practice of grouping a number of sows and their litters on a deep bed of straw has long been popular, particularly in areas where straw is relatively plentiful. In the NFU Survey, 30 per cent of farmers mixed their litters whilst still suckling.[56] It is attractive as a 'low cost' system, using simple, often general-purpose, buildings and allowing mucking out to be left to the end of the batch. An important advantage claimed for this system is that it can reduce stress. Early mixing reduces stress amongst piglets, and at weaning the batch of sows can be moved back to dry sow accommodation together without the mixing and bullying problems which result from the re-mixing of isolated sows.[57] Furthermore, if the weaners remain in their suckling accommodation until 31-36 kg, instead of being transferred to specialist rearing accommodation, the number of stressful moves to which they are subjected is reduced.

There are, however disadvantages associated with group suckling. The system is not suited to early-weaning, and implies a weaning age of perhaps 5-6 weeks. Sows may bully and fight, although this is more often a result of inadequate space or excessive group size*. Furthermore, it can be difficult to control the condition of individual sows so as to avoid thin and fat sows, and consequently poor re-breeding performance. This could be overcome by using individual feeders, although this increases capital and labour costs and reduces the attactive simplicity of this system. Cross suckling can also be a problem with large pigs getting bigger at the expense of the smaller pigs, leading to a greater variation in litter size and some reduction in overall litter weight.

Some work has been done to try and assess the comparative performance of single and multi-suckling systems, by Hillyer (1976)[61] and Petchey, Dodsworth and English (1978).[62] Some of their results are summarised in Table 20.

Hillyer failed to find any *significant* differences between single and group suckling. He did, however, note a possible trend towards a lower liveweight gain for group suckled piglets. Petchey et al confirmed the lower liveweight gain for grouped litters. This might have resulted from

---

*Recommendations for space per sow and litter (including a creep area) vary from 60 sq. ft. to 75 sq. ft.

Table 20. A COMPARISON OF SINGLE AND GROUP SUCKLING

	Hillyer		Petchey, Dodsworth & English	
	Single	Group	Single	Group
Liveweight gain (kgs) birth to weaning*	14.2	13.7	12.02	11.15
Nos. born alive	10.5	10.3	10.37	10.35
Nos. weaned	9.1	9.4	9.17	9.15
weaning-conception interval (days)	7.2	7.1	16.0	13.9
Subsequent litter size			12.77	11.35

* Hillyer's sample weaned at 8 weeks; Petchey et al, at 7 weeks.

direct stress upon the piglets, or indirectly because of stress on the sows, adversely affecting their production of milk. Petchey et al did not, incidentally, find that there was greater variation in weight within multi-suckled litters compared to litters which were single-suckled.

A second cause for concern for Hillyer was that, although mean differences in sow weight changes were comparable for both groups, there was considerably greater variation in sow weight change amongst grouped sows. This was not, however, reflected in the subsequent weaning-conception interval. A further possible effect identified by Petchey et al is a lower subsequent litter size from sows which had been suckled in groups. Unfortunately Hillyer did not report on this relationship, and Petchey et al have not established a high degree of significance.

The apparent tendency for lower liveweight gain amongst grouped piglets has to be treated cautiously. Their growth is checked when they are mixed, but they may make this up later—perhaps after weaning. Thornton comments: 'I used to find that weaners which had been grouped together were slightly lighter in weight than single-suckled litters at weaning [5 weeks], but by ten weeks of age they had made up the difference.[63] Table 21, derived from the results of Hillyer and Petchey et al, compares the live-weight gain under single and multi-suckling systems.

It can be seen from the figures that the set-back occurs at mixing (21 days in both cases), the disparity in liveweight gain closing after this until, in the case of Petchey et al, it favours the multi-suckled pigs (after weaning at 49 days). The eventual size of the disparity at 8 weeks is not large, 3.6-4.0 per cent and it is quite plausible that this would continue to diminish beyond 8 weeks, particularly where weaners which remain in multi-suckling accommodation do not suffer the stress of a move to another weaner house.

Table 21. LIVEWEIGHT GAIN UNDER SINGLE & MULTI-SUCKLING SYSTEMS
mean gain in the period (kgs)

	Birthweight (kgs)	to 21 days	21-42 days	42-56 days	birth-56 days
Hillyer					
multi-suckled	1.50	3.76	5.26	4.72	13.74
single-suckled	1.36	3.58	5.99	4.67	14.24
single > group (%)		−4.8	13.9	−1.1	3.6

	Birthweight (kgs)	to 21 days	21-28 days	28-49 days	49-56 days	birth-49 wks (weaning)	birth-56 wks
Petchey et al							
multi-suckled	1.43	3.12	0.87	6.84	2.60	10.83	13.43
single-suckled	1.49	3.19	1.47	7.08	2.23	11.74	13.97
Single > group (%)		2.2	69.0	3.5	−14.2	8.4	4.0

On the basis of these results, it appears likely that multi-suckling performance compares favourably with single-suckling, and this is certainly the case where management is of an adequate standard.

4.4 Keeping Pigs Outdoors

"Wherever possible, sows should be kept on pasture. Not only will they obtain nourishment from the grass, but the exercise which grazing out of doors provides will keep them in fit breeding condition . . . Wherever soil and climatic conditions permit, farrowing is best done in huts out-of-doors."[64]This was the advice of a leading agricultural textbook less than 20 years ago. A subsequent economic investigation of outdoor pig production carried out at Wye College and Reading University demonstrated that, largely due to lower capital and labour costs, outdoor pigs could compete very well with indoor systems.[65] Nevertheless outdoor production has lost favour with farmers. In a postal survey of 904 pig breeders in Southern England in 1967, it was found that 31 per cent of the total kept sows indoors, 7 per cent for the whole year, the remainder for part of the year only.[66] Recent surveys indicate that the proportion of holdings on which sows are kept out of doors is unlikely to be much more than 4-5 per cent,[67] [68] and the number farrowing outdoors will be considerably fewer. While the samples are not strictly comparable, the move indoors is not in dispute.

What is the reason for this trend? It is claimed that advances in indoor housing, together with the increased importance of feed costs, have eroded the competitive advantage of outdoor herds. Improvements in environmental control, improved breeds suited to indoor conditions, and earlier weaning techniques have been to the advantage of the indoor herd. Mechanisation to reduce labour costs, particularly by saving labour in waste handling, has reduced the labour needs of modern intensive systems to levels comparable with those of outdoor herds. And, as feed costs have formed an increasing part of total costs, so it has become critical to minimise feed-conversion rates.

Outdoor pig production has come to be regarded as appropriate only in special circumstances— providing a start for someone with little capital for expensive buildings, or as a break crop on favourable (i.e. light) terrain.

However, there are signs of a significant change in attitudes. A recent spate of articles in the farming press testifies to a renewal of interest in outdoor pigs.[69-73] Rapidly rising capital costs, augmented by high interest rates, together with changes in methods which appear capable of greatly improving performance, underlie this revival.

Comparisons of breeding performance for outdoor and indoor pig herds, based upon survey data, were undertaken a decade ago by Boddington (1971)[74] and Parrish (1972).[75] Their results are summarised in Table 22.

Table 22. A COMPARISON OF BREEDING PERFORMANCE AMONGST INDOOR AND OUTDOOR HERDS: 2 SURVEYS

| | Parrish (1972) | | Boddington (1968/69) | | | |
	outdoor	indoor	outdoor	Cambridge	indoor Exeter	Wye
Number of farms	7	18	41	20	27	12
pigs reared/ sow/annum	17.6	19.3	14.8	16.8	14.3	14.7
number weaned/ litter	9.0	9.0	8.3	8.6	8.1	8.8
litters/sow/yr.	1.95	2.14	1.78	1.95	1.76	1.67

Parrish's results,, from a small sample, reveal an advantage, on average, of 1.7 pigs reared per sow indoors. Boddington compared his results for outdoor herds with 3 contemporary surveys and found that the advantage to indoor pigs varied from minus half a pig to plus 2 pigs reared per sow. He rejected the Exeter and Wye comparisons as untypical of the industry as a whole, concluding that "the number weaned per sow per annum is about two lower than would be expected of most indoor herds."

It is worth examining the source of this apparent advantage to indoor systems. In Parrish's survey it is entirely attributable to the greater number of litters per sow a year (number of pigs weaned per litter is comparable), which, he tells us, is "mainly achieved by earlier weaning dates." In Boddington's survey, 70 per cent of the advantage of the Cambridge sample is attributable to the higher farrowing index. However, he does indicate that the weaning age in his sample of outdoor herds tends to be rather high, with almost 60 per cent of herds weaning at more than 6 weeks.

It appears, therefore, that outdoor herds reared fewer pigs per sow not because they were outdoors, but because they tended to wean at a later age than most indoor herds. In the 1960s and early 1970s comparatively poor breeding performance was acceptable because of the low costs involved. This is no longer the case, as an important feature of the recent revival of outdoors pig-keeping is the use of earlier weaning. Three week weaning into intensive rearing accommodation is not uncommon.

More recent survey data for outdoor pig herds is now available, published by the Meat and Livestock Commission.[76] Data from 20 outdoor herds is compared with the results for all breeding herds

Table 23. A COMPARISON OF BREEDING PERFORMANCE INDOORS AND OUTDOORS AMONGST MLC HERDS: 1979

	All herds	Outdoors	Top Third All herds	Outdoors
Number of herds	618	20	206	6
Average herd size (sows & gilts)	157	162	167	94
Litters/sow/year	2.2	2.1	2.3	2.2
Pigs/litter: total	11.2	11.0	11.3	11.4
live	10.3	10.2	10.6	10.8
Mortality of pigs born alive (%)	13.6	13.6	11.2	10.2
Pigs reared/litter	8.9	8.8	9.4	9.7
Pigs reared/sow & gilt/year*	19.1	18.5	21.7	21.7
Weight of pigs reared (kg)	19.0	19.4	19.5	22.2

*adjusted for unserved gilts

Source: MLC (1980)

recording with the MLC in Table 23. Performance is very similar for both the average and the top-third groups. Clearly outdoor herds are capable of high levels of breeding performance. Information on feed consumption from the MLC survey is reproduced in Table 24.

There is some indication that more sow feed may be used by outdoor producers, although the figures for the top-third herds suggest that this is not inevitably so. This could reflect exposure to adverse climatic conditions, or more difficult feed management. It is often argued that young pigs thrive under outdoor conditions, and the piglet weights recorded in Table 23 provide some support for this. This would offset greater consumption of sow feed.

Feed cost per £100 of gross output in Boddington's survey is broadly similar for indoor & outdoor herds. Unfortunately, it is difficult to interpret his results because outdoor herds tended to sell pigs at a later stage than most of the indoor herds. Direct comparison of feed consumption & conversion rate is therefore not very helpful.

Table 24. FEED USE IN INDOOR AND OUTDOOR BREEDING HERDS: 1979

	Average		Top Third	
	All herds	Outdoors	All herds	Outdoors
Feed/sow/year (tonne)				
sow feed	1.15	1.23	1.13	1.14
pig feed	0.50	0.52	0.59	0.63
total	1.65	1.75	1.72	1.77
Feed/pig reared (kg)				
sow feed	60	67	52	53
pig feed	26	28	27	29
Total	86	95	79	82

Source: MLC (1980b)

5 Cost and management system —
some theoretical figures

To conclude this section on pigs some hypothetical figures for
performance and costs in breeding herds are presented in Table 25.
These are proposed on the basis of the foregoing discussion and,
although they cannot be regarded as definitive in any sense, they are
intended as a plausible guide to the likely relationships between cost
and management system. It should be re-emphasised that there are
considerable gaps in the evidence upon which they are based; however,
great efforts have been made to reconcile the various sources, and
especially to make the figures consistent with MLC and Cambridge
costings. An exercise of this kind has the important advantage that it is
possible to make allowance for the extraneous factors which inevitably
confuse survey data.

It is useful to itemise a number of the more important results of the
analysis presented in Table 25. Results for both average and top-third
herds are shown in the table, and where a range is given below, this
reflects differences between these two groups.

1. Despite a slightly higher feed use than that for comparable indoor
   units, and a marginally worse breeding performance, outdoor
   systems are clearly attractive as an alternative to indoor ones. This is
   principally due to the very much lower capital costs. But an important proviso is that the results of Table 25 relate to 'suitable'
   outdoor conditions — reasonable climate and a light well-drained
   land. Such conditions are mainly found in Southern and Eastern
   England. The figures suggest, however, that outdoor pigs may also
   prove competitive even where less favourable terrain depresses
   performance.
2. If tether stalls were replaced by yards for dry sows, and flat-decks by
   verandah rearing accommodation, this would probably add between
   1 and 5 per cent to costs, as comparison between systems 1 and 2 in
   table 25 shows.
3. Replacing 3 week by 5 week weaning could add about 3-7 per cent to
   costs (see systems 1&3).
4. Late weaning, at around 7 weeks, might cost 7 per cent more than
   weaning at 3 weeks, and 4 per cent more than 5 week weaning. For
   top-third herds the disparity would be 10 per cent more than 3 week
   weaning, and 3 per cent more than 5 week weaning. (Compare 1, 3
   and 5).

5. Not using farrowing crates, as in systems 4 and 6, might add 3—5 per cent to production costs, largely due to the lower number of pigs reared and the higher labour costs.
6. Comparing the most and least intensive indoor systems, 1 and 6, reveals a difference of about 11 per cent in costs (13 per cent amongst the top third herds).

In view of these results, it is not surprising that a wide range of different systems continue to exist side by side in the pig industry. As we have shown here, intensive pig production does not have the substantial economic advantage that is commonly supposed.

NOTES TO TABLE 25

1. See Table 14.
2. See Table 23.
3. Assuming 0.2 extra pigs per litter born alive due to lower stillbirth rate.
4. Assuming loss of an extra 0.5 pigs per litter.
5. Assuming an extra 0.3 kg weight per pig reared as no accommodation change to depress growth.
6. See section on outdoor pigs.
7. Table 16.
8. 2.5% added to total sow feed to allow for extra consumption due to group feeding during lactation.
9. 5% added to total sow feed to allow for extra consumption whilst being group fed outdoors.
10. Including transport, power, litter, rent for grazing, etc.
11. Reflecting the possibility of a marginally better feed conversion in flat-decks compared to verandahs.
12. Includes an allowance of £200/sow for the initial cost of ancilliary items such as roads, passageways, fencing, waste storage, mains services, feed storage etc.

Table 25 BREEDING PERFORMANCE AND COSTS UNDER DIFFERENT SYSTEMS OF MANAGEMENT – SOME HYPOTHETICAL FIGURES

AVERAGE HERDS

Management System		1	2	3	4	5	6	7	8	9
						SYSTEM NUMBER				
Weaning age (weeks)		3	3	5	5	7	7	3	5	7
Farrowing		crates	crates	crates*	solari	crates*	solari	outdoor	outdoor	outdoor
Dry sows		tethers	yards	yards	yards	yards	outdoor	outdoor	outdoor	outdoor
Rearing		flat-deck	verandah	verandah	one-stage	one-stage	one-stage	verandah	verandah	verandah
Breeding Performance										
Farrowing index[1]		2.20	2.20	2.05	2.05	1.9	1.9	2.15	2.0	1.86
Nos. born alive		10.1	10.1	10.3	10.7[3]	10.6	10.8[3]	10.3[3]	10.6[3]	10.8[3]
Nos. reared[1]		8.8	8.8	9.0	8.9[4]	9.2	9.0[4]	9.0[2]	9.2[2]	9.3[2]
Mortality (%)		13	13	13	17	13	17	13	13	13
Weight at 8 wks (kg)		18.5	18.5	18.5	18.8[5]	18.8[5]	18.8[5]	19.0[6]	19.0[6]	19.0[6]
Nos. reared/sow p.a.		19.4	19.4	18.5	18.2	17.5	17.1	19.4	18.4	17.3
Sow feed/sow p.a. (t)[7]		1.15	1.15	1.20	1.23[8]	1.33[8]	1.37[7]	1.18[2]	1.23[2]	1.33[2]
Feed/weaner to 8 wks (kg):										
sow feed		59	59	65	68	76	80	61	67	77
pig feed		23[11]	24	21	21	18	18	26	23	20
total		82	83	86	89	94	98	87	90	97
Financial performance										
Average feed cost/t		158	159	156	155	152	152	159	157	153
Cost/weaner to 8 wks (£):										
Feed		12.98	13.18	13.41	13.83	14.33	14.89	13.87	14.10	14.88
Labour		2.58	2.84	2.97	3.57	3.14	3.80	2.58	2.71	2.89
Capital[12]		3.23	3.06	3.07	3.37	3.20	2.49	1.04	1.02	1.13
Other costs[10]		3.55	3.55	3.62	3.65	3.71	4.02	3.86	3.63	4.08
Total		22.34	22.63	23.07	24.42	24.38	25.20	21.35	21.46	22.98
Cost/kg reared (£)		1.21	1.22	1.25	1.30	1.30	1.34	1.12	1.13	1.21

* then multi-suckling

continued

Table 25 (continued)
TOP-THIRD HERDS

	SYSTEM NUMBER								
Management System	1	2	3	4	5	6	7	8	9
Weaning age (weeks)	3	3	5	5	7	7	3	5	7
Farrowing	crates	crates	crates*	solari	crates*	solari	outdoor	outdoor	outdoor
Dry sows	tethers	yards	yards	yards	yards	outdoor	outdoor	outdoor	outdoor
Rearing	flat-deck	verandah	verandah	one-stage	one-stage	one-stage	verandah	verandah	verandah
Breeding Performance									
Farrowing index[1]	2.35	2.35	2.17	2.17	2.02	2.02	2.30	2.13	1.98
Nos. born alive	10.2	10.2	10.4	10.8[3]	10.7	10.9[3]	10.4[3]	10.7[3]	10.8[3]
Nos. reared	9.2	9.2	9.4	9.3[4]	9.6	9.4[4]	9.4[2]	9.6[2]	9.7[2]
Mortality (%)	10	10	10	14	10	14	10	10	10
Weight at 8 wks (kg)	19	19	19	19.3[5]	19.3[5]	19.3[5]	19.5[6]	19.5[6]	19.5[6]
Nos. reared/sow p.a.	21.6	21.6	20.4	20.2	19.4	19.0	21.6	20.4	19.2
Sow feed/sow p.a. (t)	1.10	1.10	1.14	1.17[8]	1.27[8]	1.30[9]	1.13[2]	1.17[2]	1.27[2]
Feed/weaner to 8 wks (kg)									
sow feed	51	51	56	58	65	68	52	57	66
pig feed	22[11]	23	20	20	17.5	17.5	25	22	20
total	73	74	76	78	82.5	85.5	77	79	86
Financial performance									
Average feed cost/t	155	155	152	152	149	148	156	153	150
Cost/weaner to 8 wks (£):									
Feed	11.31	11.49	11.56	11.83	12.27	12.68	12.02	12.10	12.91
Labour	2.08	2.31	2.45	3.22	2.58	3.16	2.08	2.21	2.34
Capital[12]	2.90	2.75	2.78	3.08	2.89	2.24	0.93	1.02	0.92
Other costs[10]	2.72	3.39	3.47	3.49	3.55	3.81	3.67	3.76	3.88
Total	19.01	19.94	20.26	21.62	21.29	21.89	18.70	19.09	20.05
Cost/kg reared (£)	1.00	1.05	1.07	1.12	1.10	1.13	0.96	0.98	1.03

* multi-suckling

3 Egg production

1. Systems

In the past two decades there has been very rapid intensification of egg production. After the war, free-range and deep-litter systems were predominant but this changed dramatically in the 1960s. As Figure 10 shows, battery cages housed only 20 per cent of the laying flock in 1960 but by the end of the decade this figure had grown to 70 per cent. Only fairly recently has the pace of change abated, and about 96 per cent of the laying flock is now kept in cages.

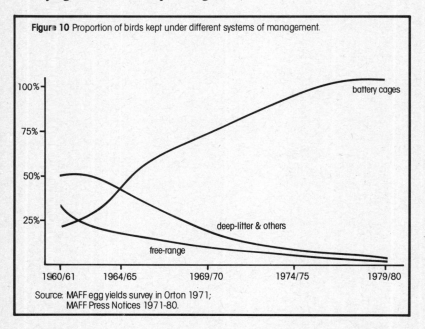

Figure 10 Proportion of birds kept under different systems of management.

Source: MAFF egg yields survey in Orton 1971;
MAFF Press Notices 1971-80.

Although the battery system now supplies most of the eggs which we consume, a number of other systems are still in use, albeit on a small scale and usually for a specialist 'free-range' market. These systems are shown diagramatically in Figure 11, and in the next section the potential

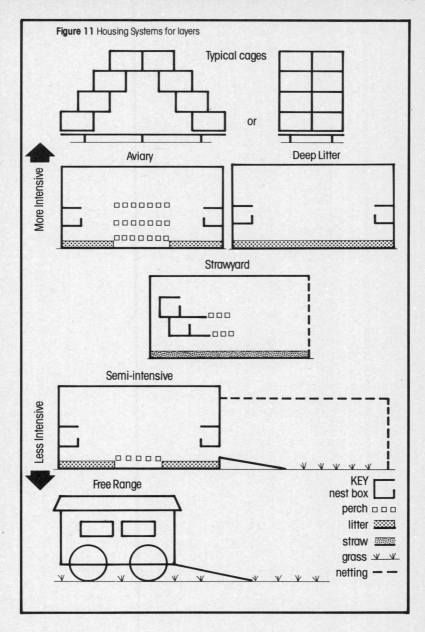

Figure 11 Housing Systems for layers

Typical cages

or

Aviary

Deep Litter

More Intensive

Strawyard

Semi-intensive

Less Intensive

Free Range

KEY
nest box
perch
litter
straw
grass
netting

of these systems, and difficulties involved in managing them, are discussed.

1.1 Free-range

Under ideal free-range conditions, birds range freely and evenly over a large area, stocked at low-densities of up to 200 per acre depending upon the quality of the land, with about 150 to the acre under average conditions. Simple, draught-free housing is provided, with roosting and nesting space, for small colonies of not more than 150 birds per house. Houses may be on sleds or wheels to enable them to be moved regularly; this is particularly desirable where the floor is slatted, as the droppings excreted while the birds are roosting (about 50 per cent of the total droppings) fall directly on the pasture. It may not be necessary to provide fencing specifically for the hens (particularly where heavier breeds are used) as they can share the pasture with other livestock and normal wire stockproof fences are often sufficient to restrain their ranging. Where houses are static, it is particularly important to keep colony size low as the area around the house receives disproportionately heavy use. Care must be taken to avoid the land becoming 'fowl sick', which may mean relocating the house every 2-3 years. Static houses have the advantage of requiring less labour and it is easier to install lighting in them to stimulate winter egg production.

Where slatted floor houses are used, less than 1 square foot (0.09m²) may be allowed per bird. However this makes it essential for them to be let out every day, preferably early in the day, if feather pecking and other vices are to be avoided. If an area of around 2 square feet (0.18m²) or more is allowed, which may be littered, this introduces a degree of flexibility in management which can prove very valuable. Birds can then be confined in adverse conditions, avoiding the loss of yield which exposure to the elements might otherwise entail. As hens lay most of their eggs early in the morning, confining them until mid-morning can greatly reduce the number of dirty eggs resulting from birds coming to the nest box straight from outdoors. However, providing more space does add to the capital cost.

1.2 Deep-litter

The deep-litter system was the first widely used method of housing layers intensively. Birds are housed on 6-9 inches of litter, ideally comprising wood-shavings or a mixture of shavings and straw. However, straw alone is frequently used and peat moss or paper shavings can be successful. This should break-down rapidly under bacterial action, creating a litter temperature sufficient to destroy or inhibit disease pathogens and parasites, and contributing to the house temperature. Clearly for this process to be successful, appropriate environmental conditions must be maintained. The litter must not be allowed to become too cold as below 10°C the bacteria which assist its breakdown are relatively inactive. The house must also not be allowed to become too humid. Particularly in winter, wet patches are liable to develop which inhibit the breakdown of the litter.

Maintaining litter quality is important for other reasons. A dry friable litter means that the birds feet will be comparatively clean, reducing the number of dirty eggs—important as egg cleaning is a costly process. Poor litter quality may contribute to outbreaks of vices such as feather-pecking and cannibalism. The causes of vices are poorly understood; breed, age, health, light, stocking density, nutritional deficiency, poor ventilation and inadequate feed or water trough space may all be responsible. Although vices may arise under all systems, they cause particular anxiety in flock systems as they can spread rapidly. The causes of vices can generally be eliminated by good management, and vigilance in removing afflicted and weak stock will usually control outbreaks. Nonetheless, the economic pressure to increase stocking density to the limit creates a perpetual hazard.

There are a number of management and design factors which influence the success of deep-litter systems. Except where the land is very free-draining, a concrete floor is advised. Although this adds to capital costs, it does allow the house to be satisfactorily cleaned and disinfected to reduce the build up of infection. Stocking density, which is generally between 1 square foot and 2.5 square feet (0.09-$0.23m^2$) per bird, has important implications. Higher stocking densities mean that it is easier to maintain house temperature with the output of body heat from the birds. However, the greater volume of excreta per unit floor area places a greater burden on the litter, particularly as the droppings contain a high proportion of moisture. To cope with this, dropping pits are provided covering 25 to 75 per cent of the floor area; the higher the stocking density, the larger is the pit required. They comprise a raised slatted area upon which the birds roost. Feed and drinking water may be located over the pits to encourage birds to spend part of the day-time on the slats. Consequently a large proportion of their droppings accumulate under the slats, which are raised sufficiently to accommodate the excreta of a full laying cycle. The area is cleaned out at the end of this period.

Insulation can be important in allowing house temperature to be maintained. Adequate ventilation to remove excess moisture is essential. Particularly where stocking densities are high, a mechanical ventilation system may be necessary to ensure that there is sufficient ventilation irrespective of external climatic conditions.

In large houses, automatic feeding is desirable. Water is provided automatically, and automatic nest boxes can be provided. The difficulty with designing automatic nest boxes is that they must be attractive nests for the birds; they may need to be littered and they must not be 'threatening', otherwise birds may prefer to lay their eggs on the floor. After the first week or so of laying, during which time the birds become accustomed to the nest boxes, the problem of eggs being laid on the floor can be overcome by designing the building interior to avoid dark corners at floor level which could prove to be attractive alternative nesting places.

It is normal practice to partition large houses, generally using wire netting. This reduces the chances of suffocation which can result from birds crowding into corners for warmth or in alarm, and contains any spread of vices. Frequently, birds are lightly 'de-beaked' to prevent damage from pecking each other.

1.3 Aviary

A recent development of the deep-litter system to provide a greater 'floor' area per bird within the same building floor area, is known as the aviary system. This seeks to make greater use of the vertical dimensions of a deep-litter house by introducing, in addition to the litter and dropping pit levels, a slatted area down the centre of the house and towards the roof. This area contains additional drinkers and feed troughs and birds can reach it by means of ladders. With the same number of birds in a house, area per bird can be increased by as much as 50 per cent over a deep-litter house; alternatively, keeping the same floor area per bird would allow an increase in the number of birds in an aviary house of 50 per cent compared to an equivalent deep-litter building. The aviary system has, so far, mainly been applied to broiler breeders and at an experimental level, although some commercial experience is accumulating. It is attractive because it offers better utilisation of space and may make it easier to maintain house temperature.

1.4 Semi-intensive

Several systems have been developed to provide a compromise between free-range and intensive deep-litter conditions. Under one system, which is referred to as semi-intensive in this report, birds are provided with a deep-litter type house from which they have controlled access to outdoor areas. Ideally two or more fenced paddocks are provided. These are used in rotation in an effort to reduce the risk of 'fowl sickness' and to ensure that green food is always available. They are stocked at the rate of 1,000-3,000 birds per acre. Access to the paddocks may be through small pop-holes to limit loss of heat from the house, and birds may be confined indoors during poor weather and until mid-morning. To ensure that the paddock remains in good condition with a continuous supply of young green shoots, it is necessary to regularly top the grass and apply appropriate conditioners. As with the deep-litter system, there is scope for mechanisation of feeding, cleaning and egg collection. The problems of litter management are similar to deep-litter although they can be exacerbated in winter when birds trail moisture in from outdoors. This can be controlled to some extent by limiting their access to outdoor areas.

Eggs produced under this system can be described as 'free-range' under the rules of the Free Range Egg Association. Several large commercial units are operated on these lines, and it is probable that a large proportion of eggs sold as 'free-range' today are produced under these conditions.

1.5 Straw-yards

Traditional hen-yards consist of a deep-litter type house with an outdoor yard of approximately twice the area of the house. The whole area is littered with straw. Despite the advantage of comparatively low-cost housing, the system has now fallen into disuse. Yields were not good, very large quantities of straw were required in the uncovered area, it was difficult to produce clean eggs from birds loose on wet straw and disease control was difficult. However, further development of the yard system has been undertaken by Dr. David Sainsbury at Cambridge University. By covering the whole area, the litter remains dry, but the structure is simple and uninsulated. This keeps capital costs low. The front can be open, although a solid front up to a height of 1 metre avoids draughts, and a simple cover during winter, perhaps using polythene sheets, may be beneficial. It is this system which is referred to here as 'straw yard'.

1.6 Battery cage systems

Battery cage systems were developed for a number of reasons. With normally 5 birds in a cage 20 inches by 18 inches which are stacked in three or four tiers, they permit very high stocking densities. The total building floor area per bird, including passageways between cage batteries, may be no more than 0.66 square feet ($0.06m^2$). This minimises the cost of the building and other fixed equipment per bird (although the cage cost offsets this to some extent) and enables a minimum temperature of 21°C, regarded as commercially optimum, to be maintained more easily without resort to artificial heating. The system is well suited to mechanisation, with eggs collected by conveyor belt, feed delivered by automatic hopper, and excreta either scraped or conveyed automatically, or allowed to fall into a deep-pit below the cage area for removal at the end of the laying cycle. Houses are generally well insulated with a closely controlled environment.

2. Performance and Production Costs

We turn now to consideration of the costs and performance likely to be achieved under each system. To assist in the investigation data was collected from 25 non-battery enterprises, ranging from 200 to 15,000 layers and comprising seven deep-litter, seven semi-intensive and eleven free-range enterprises. This is not a large sample, and in a number of cases the data was imperfect, generally as a result of poor farm records. It has, nonetheless, provided useful up-to-date information on, in particular, egg yield and labour use on a number of less intensive farms. Information was gathered principally by farm visits, supplemented by telephone interviews and written questionnaires. The farmers were identified mainly through personal contacts and press articles. The size distribution of the holdings is given in Table 26.

Table 26. SIZE DISTRIBUTION OF ENTERPRISES FROM WHICH DATA WAS COLLECTED—EARTH RESOURCES RESEARCH SURVEY

No. of layers	No. of enterprises
200-499	6
500-999	6
1,000-1,999	4
2,000-5,000	5
over 5,000	4
	25

It is important to bear in mind that in comparing different systems it is costs and performance on modern, commercial-scale enterprises which are at issue. In what follows egg yield and hen mortality are considered and the three main cost areas— labour, feed and capital— are examined.

2.1 Egg Yield

Many environmental factors influence egg yield, but the two which are of most significance here are light and temperature.

It is well known that the reproductive activity of the hen is closely associated with light patterns, which are in turn associated with the seasons. Increasing day-length stimulates sexual maturity and reproductive activity; decreasing day-length has the reverse effect. This control over lighting makes it possible to avoid the loss of production which occurs in conditions of declining natural light. It also makes it possible for peak production and moult to be scheduled independently of season, avoiding the pattern of glut and shortage which can arise where the environment is less closely controlled. One trial indicated that where hens were kept under free-range conditions between August 1965 and January 1966, the percentage egg production* was 60 per cent where the birds were lit, compared to 38 per cent where there was no artificial light.[2] However, historical examples such as this may exaggerate the effect of light on egg production as modern fowls are less sensitive to day length. Turning to the effect of temperature upon egg production, hen-housed egg production increases by about one egg per bird per year for each degree the temperature rises, up to about 25 to 30°C. Thus layers kept at 15°C would each produce 6 eggs fewer a year than those kept at the 21°C which is currently regarded as the commercial optimum. However, strain, stocking density, acclimatisation and feather cover will all affect this relationship and cannot be assumed to be independent of the management system.

Experimental Results. A number of trials have been carried out in several countries to compare the egg yields obtained under various systems of management. Comparison of deep-litter and cage systems has received particular attention in North America and in most cases litter systems did as well as or rather better than cages.[5-9] However it is not easy to draw out the implications of these results for UK commercial conditions. The results are dated and refer to climatic conditions rather different from those in the UK, which will affect ease of management— particularly of litter which is more difficult to manage in humid UK conditions. In most studies stocking density was low, generally 3 square feet (0.28 square metres) or more on the floor, with cage birds kept singly. This is rather different from modern commercial practice.

In the UK, Bareham (1972) reported yields 20-30 per cent lower on litter than in cages.[10] The sample was, however, small and single-bird cages were used. Comparisons being made currently at the National Agricultural Centre have favoured cages, since there have been a high proportion of cracks and second-quality eggs on litter. However, their

*i.e. actual egg production as a percentage of 1 egg per bird per day

deep-litter trial is being held in a recently converted building and there have been considerable problems in managing the litter, at least in the early stages.

The Swiss Foundation for the Promotion of Poultry Breeding and Keeping undertake regular laying performance trials,[11] including comparison of cages with litter. The averaged results for the 3 years 1976-80 marginally favoured the litter units. However, although numbers involved were quite large, the results must be interpreted with caution because the trial was set up to compare breeds rather than two commercial systems.

Semi-intensive systems, with a grass run, have been compared with single-bird cages and deep-litter.[13] Egg production was in the region of 15-20 per cent lower under these systems. Unfortunately it is not made clear in either case whether the semi-intensive houses were lit or not, which clearly has an important bearing on the results.

Experimental work at the Celle Institute in West Germany is in progress to compare cage, deep-litter and semi-intensive systems. Three-tier, four-bird cages, with 0.05 square metres per bird, are being compared to a system in which birds are kept on a half-litter, half-slatted area at a rate of 0.17 square metres (1.8 square feet) per bird. Both areas are mechanically ventilated and windowless. Another group of birds is being kept in a naturally ventilated house with windows and access to pasture stocked at 500 birds per acre. All housing systems are lit for 16 hours daily.[14] Final results are not available, although preliminary results suggest that hen-day egg production is similar in all 3 systems. However, it should be noted that, due to a high proportion of passage space, house-stocking densities in the cage house are below commercial stocking rates, resulting in comparatively low winter temperatures, which reached a minimum of 6°C during the first winter.[15] It is possible that this has adversely affected rate of lay in cages.

Egg production from birds in multi-bird cages in a controlled environment house has been compared to results from a simple, fully covered, straw yard by Dr. Sainsbury at Cambridge University. Results for a five year period revealed that the birds in straw-yards yielded as well as those in cages.[16] However, Dr Sainsbury does not indicate how far the cage house reflects commercial conditions.

Drawing conclusions from these experimental results is far from easy. They cover a wide range of factors and circumstances over quite a number of years, during which overall performance levels have shown considerable change. Some of the results appear to conflict. Difficulties of interpretation are compounded in some instances by the lack of crucial information about lighting and house temperatures. In general, the impression is of more variable results on litter, perhaps due to more exacting management requirements. A short experiment, particularly if operated by staff unfamiliar with litter management, may not allow time for any problems arising to be overcome. Lighting is clearly of very great importance. Semi-intensive systems have been shown to be capable of yields comparable to intensive systems where lighting is provided, although it is not certain that these results could be regularly repeated. Stocking density also appears to be important and it is possible that the performance of flock-housed systems is rather more vulnerable to higher stocking densities than in the case of cages. Breed and strain differences are likely to further influence comparative performance. Before drawing any further conclusions, it will be useful to consider the various survey results which are available.

Survey Results. Since the war, MAFF has undertaken a regular Egg Yield Survey. Based on a changing sample of 350 holdings selected each month, information is collected by postal returns and contact by visiting officers, and the results are classified according to the system of management. The results for recent years are given in Figure 12. While the less-intensive systems show more variable results than cages, yields have increased at a very similar rate under all systems. This is confirmed by Table 27.

Table 27. EGG YIELD PER BIRD AS A PROPORTION OF BATTERY EGG PRODUCTION: ROLLING 5 YEAR AVERAGE (%)

5 years ending:	free-range	deep-litter & others
1973/4	77.9	88.4
1974/5	77.0	88.2
1975/6	76.9	88.6
1976/7	76.4	89.0
1977/8	76.2	89.0
1978/9	77.1	90.0
1979/80	77.8	90.7

Source: MAFF Egg Yield Survey

It appears from these survey results that egg yields per bird per annum are something like:

battery	250
deep-litter & others	225-30
free-range	195

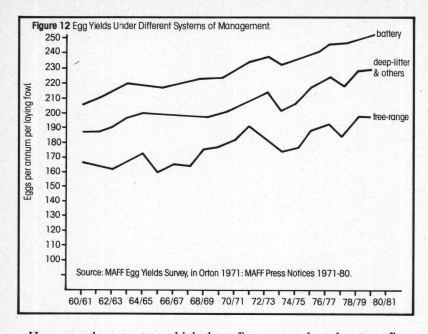

Figure 12 Egg Yields Under Different Systems of Management.

Source: MAFF Egg Yields Survey, in Orton 1971: MAFF Press Notices 1971-80.

However, the extent to which these figures may be taken to reflect results obtainable under normal commercial conditions is very much in doubt. In general, the sample of 350 farms contains 70 free-range enterprises, 70 deep-litter 'and others', and the remainder are battery units. Finding sufficient commercial battery units is not too difficult. However, in order to find a changing sample of around 140 non-cage systems, it has been necessary to include flocks of as few as 10 birds, as there are nowhere near sufficient commercial-scale non-battery units to satisfy the sampling requirements. No information is given about the average age of birds in the flocks sampled (laying performance deteriorates after the first year in lay), stocking densities, or whether deep-litter houses have a controlled environment. Furthermore, deep-litter units are lumped together with an unspecified range of systems known as 'and others', which presumably includes a considerable variety of semi-intensive, straw yards and other types of management. There is a tendency, therefore, for the survey to compare large commercial battery flocks with small 'back-yard' less-intensive flocks. The latter sometimes includes some holdings with fewer than 100 layers of indeterminate age kept in makeshift unlit buildings and managed by someone with little experience of poultry keeping. This severely limits the value of the comparison.

In 1978, the Swiss Centre for Poultry collected data on the performance of 65,000 layers in 38 flocks, two-fifths of which were on litter, the remainder in cages.[17] The results, together with those for a similar exercise in 1976 and 1977 are given in Table 28.

Table 28. A COMPARISON OF LAYING PERFORMANCE IN DEEP-LITTER AND CAGED FLOCKS IN SWITZERLAND

Mean laying performance for 14 months in lay (%)

	1976	1977	1978
Battery	72.9	73.9	73.9
Deep-litter	74.3	76.3	72.8

Source: Swiss Centre for Poultry, 1978.

There appears to be no significant difference between the results achieved under the two systems. No details are given of the conditions under the two systems, but the high average rate of lay suggests that they are likely to be close to the optimum.

The yields obtained by enterprises in our survey are set out in Table 29.

Table 29. EGG-YIELD PER BIRD UNDER DIFFERENT SYSTEMS—EARTH RESOURCES RESEARCH SURVEY (1980)*

Hen-housed average: (eggs per bird per annum)	deep-litter	semi-intensive	free-range
< 200	1	1	1
200-220	0	1	1
221-240	1	6	2
241-260	1	2	2
> 260	4	1	0

* information obtained from 24 farms.

It should be noted that the results from the two highest yielding free-range units were rather less well documented than most of the other figures and should therefore be treated with particular reserve. With this in mind, it is possible to see a pattern emerging. The tendency is for birds kept on deep-litter to yield rather better than under other systems, with the majority of holdings achieving a hen-housed average in excess of 260 eggs, a good result by any standards.

The performance of semi-intensive and free-range systems is less easy to evaluate. There is a tendency, however, for semi-intensive units to lie toward the top of the range, and free-range units towards the lower end. Because of the small sample size, it did not prove useful to distinguish between lit and unlit free-range systems, or between the various types of semi-intensive system.

From a survey of 70 battery egg producers in 1976/77, Burton (1978) found an average yield of 242 eggs per bird, very close to the results of the MAFF Egg Yield Survey for that period.[18] He shows that specialist egg farms, large farms, those with full environmental control, and farms keeping the laying cycle to less than 58 weeks are all associated with comparatively good egg yields. This suggests a yield somewhat above 242 eggs per bird for a good commercial unit, although perhaps not by much as the variations around the average were not large.

Drawing together all of these results, the following estimate of egg yield under different systems of management appear plausible, given good commercial conditions.

Table 30. ESTIMATES OF HEN-HOUSED AVERAGE EGG YIELD PER BIRD PER ANNUM

Battery cages	260
Deep-litter	250
Aviary	245
Straw-yard	245
Semi-intensive	235
Free-range	225

2.2 Second quality eggs

When an egg is laid in a cage, it quickly rolls away on the sloping wire floor to await collection out of reach of the bird. There is little time for eggs from caged birds to become soiled, damaged or eaten by the birds. Further, the separation of birds from their droppings means that cage floors and birds feet are relatively clean. The incidence of dirty eggs is low, which is important since, as has been indicated, egg cleaning is a time-consuming process.

Where birds are kept loose-housed and eggs are laid in nest-boxes, a number of problems can arise. In the first place, birds may not choose to lay in the nest-boxes and eggs laid elsewhere are vulnerable to breakage and take time to collect. This often happens as birds come into lay before they become accustomed to nesting. However, this can largely be eliminated by good design and management. Nest-boxes should be made attractive to nesting birds. They need to be secluded, of the right dimensions, and provided with nesting material. Alternative nesting sites, in dark corners for example, ought to be eliminated.

Similarly, good management can greatly reduce the number of cracked and dirty eggs. Regular egg collection and the provision of plenty of nest-boxes reduces congestion and the risk of damage. Nest boxes have the advantage of generally being littered, unlike wire cage floors which can result in cracked eggs, particularly where shells are weak. Dirty eggs in nest boxes are largely caused by birds entering the boxes with wet and dirty feet. There are many ways of combatting this. Good litter management should ensure a clean, dry surface to the litter. If the approach to nest boxes is slatted this will clean birds' feet. Where range is provided, keeping birds indoors until mid-morning when most eggs have been laid (and collecting soon after) will reduce the damage from muck trailed into the house from outside. Where there is sufficient space indoors, birds need not be let out at all in wet weather.

Where dirty eggs do arise, which happens under all management systems to some degree, they can still be dry-cleaned and sold. Allowance for time taken to do this has already been made in the earlier estimates of labour requirements. To allow for the possibility of more cracked eggs under loose-housed systems compared to cages, the earlier estimates of non-cage egg yields in Table 30 have been reduced by a notional 2 per cent in the final costings, which appear in Table 39. This is in addition to eggs which are lost altogether, and which therefore do not appear in the figures for egg yield.

2.3 Mortality

There is not a priori case for expecting mortality to be higher under one system than another as each has advantages and disadvantages. Cages separate the stock from their excrement, greatly reducing the incidence of disease. Disease need not, however, be a problem under litter systems, providing that the litter is well-managed so that it maintains a temperature and moisture content which inhibits the growth of pathogens and parasites. However litter systems are clearly more susceptible to deficiencies in management. Similarly, the small colony size in cages can limit the spread of vices, which can spread rapidly where birds are housed in flocks. Conversely, cages provide little opportunity for bullied birds to escape serious damage, and the degree of confinement and restriction of behaviour patterns may lead to vices arising with greater frequency. Caged birds are also less able to regulate their micro-climate, leaving them vulnerable to poor management or malfunctioning equipment.

Permitting birds access to outdoors provides greater opportunity for them to vary and balance their own diet, gives access to fresh air and sunlight, and more opportunity for exercise. Vices may be less of a problem where birds have a more varied environment, and where the comparatively low-stocking densities make it easier to avoid aggressive encounters with other birds. Exposure to a more varied climate enhances a bird's adaptability to changes in conditions, by encouraging a denser growth of feathers, for example. However, birds will be exposed to adverse weather which, as well as any direct affect upon their health, can make litter management more difficult. There is also the risk of poorly-managed pasture becoming 'fowl sick', and harbouring parasites. In addition birds outdoors are susceptible to predator attacks, particularly from foxes in the UK, and this can be a serious problem.

A number of the experimental projects which provide information on egg yield under different systems also give data on mortality. Gowe (1956),[19] Lowry et al (1956)[20] and Logan (1965)[21] all report higher mortality on deep litter in trials carried out in N. America, while Bailey et al (1959)[22] found comparable results. In the UK, Bareham (1972)[23] found a similar level of mortality in one group, with rather higher mortality in a second group on litter. The Swiss Laying Performance Trials compared mortality on the floor and in cages which, over the years 1968-78, favoured cages, although by a fairly small margin.[24] Coles (1960)[25] found slightly lower mortality amongst birds kept semi-intensively, compared to deep-litter. Preliminary results from the Celle Institute suggest slightly higher mortality under litter and semi-intensive conditions, attributable to vice problems and predation respectively.[26]

Some of the problems with interpreting these results have already been noted. The tendency to use single-bird cages in the earlier trials, and the generally low stocking densities are particularly confusing.

The Earth Resources Research survey did not provide sufficient information on mortality to be useful in making a comparison. One point which emerged from the farm visits was that a number of farmers considered light debeaking to be an important management tool for loose-housed birds. The MAFF Egg Yields Survey does not include data on mortality. However, mortality is included in the data collected by the Swiss Centre for Poultry and is very similar under cage and deep-litter conditions.

This evidence is insufficient to make any firm prediction of mortality under different systems of management. The most plausible conclusion appears to be that mortality is likely to be more variable in flock systems, and may be equal to or rather higher than in cages. Where greater mortality occurs amongst loose-housed birds, this is likely to result from poor litter management, vice problems, or predation where birds are outdoors.

2.4 Feed consumption

Feed use is affected by temperature, egg production, wastage and the availability of other sources of nutrients. Each of these factors will be different under the various systems of management.

Temperature. It is well established that as temperature rises so feed intake falls. At higher temperatures birds lose less heat to their environment and this reduces their dietary energy needs.

A minimum temperature of 21°C is widely regarded as the commercial optimum and can normally be maintained in densely stocked cage houses without artificial heating. Good insulation and carefully regulated ventilation ensure that the metabolic heat from the birds is sufficient to maintain 21°C.

The importance of the thermal properties of buildings is shown in the graph in Figure 13. Where no protection is provided, so that the difference between the internal and the external temperature is zero, feed consumption is 7-8 per cent more than where the building is

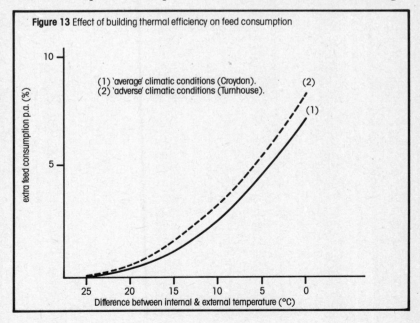

Figure 13 Effect of building thermal efficiency on feed consumption

(1) 'average' climatic conditions (Croydon).
(2) 'adverse' climatic conditions (Turnhouse).

extra feed consumption p.a. (%)

Difference between internal & external temperature (°C)

capable of maintaining an internal temperature 20°C above the temperature outdoors. Where there is a temperature gradient* of 20°C, feed consumption is largely independent of the external temperature because the internal temperature can be kept close to the optimum for most of the time.

Egg production. Layers use feed to produce eggs and therefore have higher dietary needs the more eggs that they produce. It has been estimated that this amounts to approximately 0.5 kgs of feed per 10 eggs. [27] As we have seen, egg production is likely to vary between systems and so this factor will need to be taken into account in estimating feed consumption.

Wastage. Vermin and, especially, wild birds can be a problem where they have access to feed troughs. This is particularly the case under extensive systems. Hewson (1979) assumes that wastage adds about 6 per cent to feed use under extensive systems, while admitting that this assumption is an arbitrary one. [28] There is, unfortunately, no information on wastage to permit a better estimate to be made.

Contribution from the pasture. Where there is access to pasture this can be expected to contribute insects, seeds and a little grass to the hens' diet, particularly during the warmer months of the year. This contribution will be influenced by stocking density and grassland management and may favour extensively kept birds quite significantly. [29-31] Unfortunately most comparative information is very out of date and so does not take account of changes in diet and feed efficiency over the past decade or so.

Developing a model. For this study a simple model was constructed to make estimates of feed consumption under different systems of management. This model takes account of the thermal properties of the housing, stocking density, the amount of time that layers spend indoors, and their egg yield. Details are given in Appendix 2.

The results, which take no account of wastage or of the nutritional contribution from the pasture, are presented in Table 31. In what follows these results are considered in the light of survey and experimental data to produce final estimates.

Survey and experimental data on feed consumption. Information collected in the Earth Resources Research farm survey is summarised in Table 32. This must be treated very cautiously as not only is the sample

*i.e. the difference between the temperature inside a building and the external temperature.

Table 31. PRELIMINARY ESTIMATES OF FEED CONSUMPTION UNDER DIFFERENT SYSTEMS OF MANAGEMENT (KG PER BIRD P.A.)

battery	deep-litter 1.5ft²/bird	2.5ft²/bird	aviary (2ft²/bird)	straw-yard	semi-intensive	free-range
43.0	42.6	42.8	42.3	44.4	43.4	43.9

Table 32. FEED CONSUMPTION—EARTH RESOURCES RESEARCH SURVEY (1980)

Feed consumption gms/bird/day	deep-litter	semi-intensive	free-range
110-120	2	4	2
121-130	2	2	1
131-140	−	1	−
141-150	−	1	1
>150	−	−	2

(information obtained from 18 enterprises).

very small, but it is based largely upon farmers' estimates rather than detailed records.

Taking battery feed consumption to be about 118g per day, it appears that the deep-litter and semi-intensive units compare reasonably favourably. It is interesting to see the considerable variation in the results of the free-range producers. While some compare favourably, others record very high feed consumption. This is likely to reflect considerable variations in the amount of wastage and, to some extent, differences in the quality of pasture.

The laying performance trials carried out by the Swiss Foundation for the Promotion of Poultry Breeding and Keeping, and the survey carried out by the Swiss Centre for Poultry (both described earlier), compare cage and litter systems and include details of feed consumption. These are reproduced in Table 33, from which it can be seen that feed consumption is 4-7 per cent higher on litter under these conditions.

Table 33. FEED CONSUMPTION UNDER LITTER AND CAGE CONDITIONS (GMS PER BIRD PER DAY)—SOME SWISS RESULTS (1978)

	Swiss Centre for Poultry [1]	Swiss Foundation for Poultry [2] Breeding
litter	123.5	125.7
cage	115.4	120.4

[1] average 1-14 months in lay.   [2] brown eggs.

Results reported by Dr. Sainsbury for his straw-yard system at Cambridge[32]indicate that over a 5 year period feed consumption in the straw-yard is very similar to that of a matched group in cages. It is not, however, clear whether the performance in cages is as good as would be expected under commercial conditions.

Preliminary results from work at the Celle Institute in West Germany indicate feed intake of 124g per bird per day in cages; 129g per bird per day on deep-litter; and 131g per bird per day in a semi-intensive system.[33] The cage results may be rather high as the houses are less densely stocked than under commercial conditions.

It is clear that with the present state of knowledge there can be no definitive estimate of the relationship between feed use and management system. However, on the basis of the information reviewed above, estimates can be made and appear in Table 34.

Table 34. FINAL ESTIMATE OF FEED CONSUMPTION UNDER DIFFERENT SYSTEMS OF MANAGEMENT

	kg of feed consumed per bird p.a.	gms of feed consumed per egg produced
cages	43.0	165
deep-litter: 1.5ft²/bird	42.6	170
2.5ft²/bird	42.8	171
aviary	43.0	175
straw-yard	45.5	186
semi-intensive	46.0	196
free-range	44.0	196

Feed consumption per bird is almost identical under cage, deep-litter and aviary conditions, although feed per egg is lowest in cages because of the higher egg yield. Free-range birds have a lower feed consumption (per bird) than birds in semi-intensive and straw-yard systems, partly because of lower egg yields and partly due to the contribution from the pasture.

2.5 Capital costs

The difficulties associated with estimating capital costs are discussed in the section on pig systems. Briefly, differences in supplier, farm situation and standard of construction lead to large variations in costs in practice. The following estimates in Table 35 are a reasonable guide and are based largely upon ADAS estimates.

Table 35. CAPITAL COSTS PER BIRD—MID 1981 PRICES (£)

	Buildings	Equipment	Total
Cages	2.20	3.30	5.50
deep-litter			
1.5ft²/bird	5.50	2.00	7.50
2.5ft²/bird	8.20	2.50	10.70
Aviary			
0.5ft /bird	5.00	2.30	7.30
straw-yard	5.50	2.00	7.50
semi-intensive	6.00	2.50	8.50
free-range	6.00	2.00	8.00

In Table 36 buildings are depreciated over 15 years and equipment over 10 years to give the following table of annual capital costs per bird.

Table 36. ANNUAL CAPITAL CHARGES PER BIRD (£)*

Real Annual interest	3%	5%
cages	0.59	0.67
deep-litter 1.5ft²/bird	0.72	0.82
2.5ft²/bird	1.02	1.09
aviary	0.72	0.82
straw-yard	0.72	0.82
semi-intensive	0.82	0.94
free-range	0.76	0.87

* includes allowance for 3 weeks depopulation of housing per annum.

Burton (1981) assumes rather lower capital costs for the non-cage systems, as he claims that there would be some scope for adaption and conversion of cage houses to less intensive use.[34] While this is undoubtedly true, and an important element in considering the cost of transition to less-intensive systems, it should not be included directly in a comparison of production costs, where the question at issue is comparative use of resources rather than the effects of a ban on cages. Furthermore, the opportunity for conversion would only be available to some producers and largely on a one-off basis; future investment would be predominantly in purpose-built accommodation.

In his estimates of capital costs, Morris (1981) assumes comparatively high non-cage capital costs, particularly for deep-litter, on the basis that considerable investment would be made in labour-saving equipment such as automatic egg collection.[35] However, this is an arbitrary assumption; the technologies involved are unproven and without knowing the cost of developing effective equipment it is not possible to estimate either the cost involved or whether it might prove attractive commercially. Consequently, the present estimates provide only for those technologies, which have proved commercially successful, such as automatic feeding and monorail egg collection.

Battery cages have the lowest capital requirement per bird because the buildings are so densely stocked. Deep-litter with 1.5 square feet per bird, aviary accommodation and straw-yards are the cheapest alternatives. Deep-litter and aviary keep costs down by adopting comparatively dense house stocking; straw-yards, while allowing birds more space, use cheaper buildings. Where birds are permitted to range, some capital is required for external equipment, and the small size of free-range housing increases the cost per bird. Low-density deep-litter housing is the most expensive as the comparatively high costs of providing a controlled environment is spread over relatively few birds.

2.6 Labour Costs

The difficulties involved in estimating labour requirements are well known. Differences between systems are clouded by the large differences within systems, depending upon, for example, labour quality and motivation, design and layout of buildings, and the availability of labour-saving devices. Care must be taken too, to ensure that like is compared with like; farms marketing their eggs at the gate will need extra labour, compared to a farm where the eggs are collected by a packing station. Where there are several enterprises on a farm, the difficulty of accurately dividing labour time between each enterprise is notorious. Even on specialist poultry farms, some farmers may be rearing their own replacements, which needs to be distinguished from time involved in egg production. One way of tackling this problem is to stand with a stop-watch and time individual activities, which can then be aggregated. This approach was adopted by Sturrock (1960), and his results for deep-litter and free-range systems are presented in Table 37.[36]

Interpreting these results presents some difficulties, particularly as Sturrock provides no further details about the holdings. Moreover the methodology itself has weaknesses, as it does not measure the actual number of birds managed by one person. Conditions on a well-planned and efficient holding can differ greatly from the 'average'; for deep-litter, the number of birds managed by one person in good conditions is twice the average, but Sturrock only provides average results for the extensive systems.

Table 37. LABOUR USED PER 100 BIRDS (MINUTES PER WEEK)

| | Deep-Litter | | Free-Range |
	Average	Good	Average
Feeding	35	21	49
Watering	2	1	32
Cleaning	34	9	38
Egg Handling	75	38	95
Other Work	9	9	63
Total	155	78	277
Birds per man	2,100	4,200	1,200
No. of flocks		12	10

Source: Sturrock (1960)

Furthermore, in his sample, watering takes a substantial time in the extensive systems. However, it is possible to largely eliminate this, certainly for static houses, by supplying automatic drinkers which are mains fed or gravity fed from large infrequently filled tanks near or attached to the house. Similarly, ad lib feed hoppers which require filling perhaps only twice a week could substantially save labour. At the time of Sturrock's study wages were comparatively low and margins good so that the incentive to design systems which minimised labour use was rather less than now. Furthermore, as the vast majority of flocks were small and non-specialist, the opportunity to take advantage of scale economies must have been very limited.

However, there are factors operating in the opposite direction. Sturrock calculated on the basis of a 54 hour week. Hours worked are now closer to 46,[37] some 15 per cent less. The deep-litter houses studied did not have mechanised feed handling and were small by modern standards. Sturrock describes a 'large' house of only 1000 birds. The hen yards fare less well than deep-litter, partly because they were mostly converted buildings. Also, the amount of egg-cleaning is normally greater than for litter.

The Earth Resources farm survey results are presented in Figure 14 (21 enterprises provided usable information on labour requirements.) These figures were arrived at by taking the total hours spent on the unit, excluding chick rearing and egg marketing, and calculating the number of birds which could be managed by one person working an average working week of 46 hours. This is, of necessity, a crude method. A particular difficulty arose in extrapolating the results of smaller enterprises as this compounds any errors and fails to take account of scale economies.

Figure 14 Number of birds managed by one person – survey results.

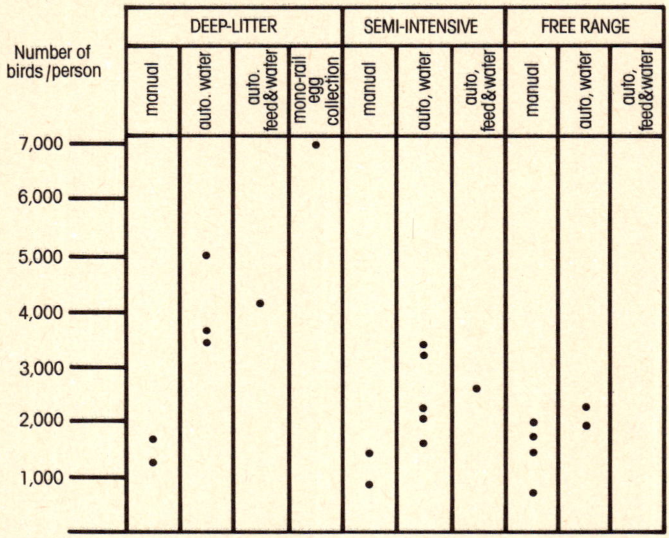

Number of birds/person	DEEP-LITTER				SEMI-INTENSIVE			FREE RANGE		
	manual	auto. water	auto. feed&water	mono-rail egg collection	manual	auto. water	auto. feed&water	manual	auto. water	auto. feed&water
7,000				•						
6,000										
5,000		•								
4,000			•							
3,000		••				••	•			
2,000	•					••			•	•
1,000	•				• •	•			•	

On the basis of these results, the work of Sturrock,[38] and other estimates for deep-litter[39] and extensive systems,[40] the estimates in Table 38 were made.

These figures are intended to reflect good commercial conditions. They do not, however, represent best practice, which could be substantially higher.

No information is available on labour use under aviary conditions. By permitting more birds to be housed in a deep-litter building, it is likely that an aviary system would allow more birds to be managed by one person, providing that additional difficulties in managing the stock and working in more crowded conditions did not offset this.

3. Production cost comparisons

Taking the estimates which have been developed earlier in this section, comparative costings are presented fully in Tables 39 and 40. Battery cages and intensive flock systems, ie. deep-litter and aviary, show very similar costs, while straw-yard, semi-intensive and free-range systems are progressively more costly. The extent of the cost disparity is

Table 38. NUMBER OF BIRDS WHICH CAN BE MANAGED BY ONE PERSON

Battery cage:	
automatic feed, water and egg collection	10,000*
Deep-litter:	
basic system	3,000
with automatic water	4,000
with automatic feed and water	5,000
with monorail egg collection	6,000
with automatic egg collection	8,000
Straw yards:	
basic system	2,500
with automatic water	3,000
with automatic feed and water	4,000
with monorail egg collection	4,500
Semi-intensive:	
basic system	2,000
with automatic water	2,500
with automatic feed and water	3,000
with monorail egg collection	3,500
Free-range:	
basic system	1,500
with automatic water	1,800

* following Morris (1981)[41] and MAFF (1981)[42]

influenced significantly by the density at which cages are stocked. The effect of encouraging or enforcing lower cage stocking densities would be to make non-cage systems more attractive. At 3 birds per cage, straw-yard would not be unattractive as alternatives to cages, although significantly less attractive than the intensive litter systems.

The reasons for the difference in battery and non-battery costs are explored in Table 41.

Lower egg yield is clearly the most important cause of increased production costs, especially for the intensive flock systems. Labour costs are the other major factor, becoming more important the less intensive the system. In contrast, capital costs are a significant element in the extra cost of eggs produced in the intensive flock systems, but are less important amongst the more extensive systems. It can be seen that the extra cost of land under extensive conditions is insignificant.

In Table 42 our costings are compared with several other estimates which have been made recently. Our estimates are the lowest for deep-litter and, with the exception of Morris, for free-range also. All of the estimates for straw-yards are very similar.

Table 39. PERFORMANCE DATA – 1980/81

	Battery-cages	Deep-Litter¹	Aviary	Straw-yard	Semi-intensive	Range
PHYSICAL						
Egg yield per bird (HHA)²		245	240	240	230	220
Birds per person	10,000	8,000	9,000	4,500	3,500	1,800
Feed consumption per bird (kg/pa)	43.0	42.6	43.0	45.5	46.0	44.0
FINANCIAL (£) (1980/81 prices)						
Annual capital cost per bird³	0.67	0.82	0.82	0.82	0.94	0.87
Labour cost per bird⁴	0.45	0.56	0.50	1.00	1.29	2.50
Feed cost per bird⁵	6.02	5.96	6.02	6.37	6.44	6.16
Livestock depreciation	1.72	1.72	1.72	1.72	1.72	1.72
Land⁶	–	–	–	–	0.02	0.20
Other costs	0.30	0.30	0.30	0.30	0.30	0.30
Total cost per bird	9.16	9.36	9.36	10.21	10.71	11.75
Cost per dozen eggs (to achieve a margin of 50p per bird p.a.)	0.45	0.48	0.49	0.54	0.58	0.67

Notes
1. 1.5ft²/bird.
2. Hen-housed average, after allowing for second quality eggs.
3. At 5% real interest p.a.
4. At a labour cost of £4,500/person/year.
5. At £140/tonne.
6. At £30/acre; free-range, 150 birds/acre; semi-intensive, 1500 birds/acre.

Table 40. DIFFERENCES IN PRODUCTION COSTS BETWEEN EXTENSIVE SYSTEMS AND CAGES STOCKED AT DIFFERENT DENSITIES (PER CENT GREATER THAN BATTERIES)

	5 birds per cage	4 birds per cage*	3 birds per cage*
Deep-litter	7	1	−9
Aviary	9	3	−7
Straw-yard	20	14	3
Semi-intensive	29	22	10
Range	49	41	28

* using estimates for 4 bird and 3 bird cages derived from ADAS evidence [43]

Table 41. PERCENTAGE OF THE EXCESS OF NON-BATTERY OVER BATTERY PRODUCTION COSTS ATTRIBUTABLE TO DIFFERENT FACTORS*

	Deep-litter	Aviary	Straw-yard	Semi-intensive	Range
Lower egg yield	83	88	50	50	52
Higher feed cost	−5	−	17	13	2
labour costs	9	3	26	27	38
capital cost	13	9	7	9	4
land cost	−	−	−	1	4

* columns do not total 100% as figures have been rounded.

Table 42. VARIOUS ESTIMATES OF THE EXTRA COSTS OF NON-CAGE OVER CAGE SYSTEMS

	Carnell	ADAS[44]	Burton[45]	Morris[46]
Deep-litter	7	17	22	17
Straw-yard	20	17	−	19
Semi-intensive	29	−	−	−
Free-range	49	79	99	43

Within these overall cost differences there are a number of other differences in the composition of the costings. Morris assumes comparatively high egg yield and low labour use for non-cage systems, but this is offset by high capital costs. Both Burton and ADAS calculate high free-range costs. Both envisage labour costs some 10 times greater than for indoor systems (approaching twice our assumption) and extra feed costs contribute too, especially in Burton's analysis. For deep-litter, ADAS and Burton assume comparatively high labour requirements, while Morris assumes high capital costs, offset by a high egg yield.

An important difference between our analysis and the other costings is that nobody else considers a semi-intensive system. This system includes an element of outdoor range and probably supplies a large part of today's "free-range" egg market. It also has significantly lower production costs than a true free-range system.

4. The retail price of eggs

4.1 Distribution margins

It is often simply supposed that distribution costs are a proportion of production costs. However, it is difficult to see why the fact that eggs from one enterprise cost 50 per cent more to produce than those from another enterprise implies that they should cost 50 per cent more to distribute as well. Other factors being equal, distribution costs should be equal. Of course, other factors may not be the same and, in particular, the size of the enterprise may be different. It has been argued that non-cage enterprises would tend to be smaller than existing battery farms and that this would entail higher transport and other costs.

An existing battery unit converted to non-cage production would mean that existing buildings would hold fewer birds, and higher labour requirements would limit the number of birds which could be managed by the existing labour force. If egg output were to be maintained, more labour and buildings would have to be made available for egg production. In some cases, the expansion of existing units would be constrained by a lack of capital, land or suitable labour, or by institutional factors such as difficulty in getting planning permission. The more extensive the system adopted, the more severe these constraints are likely to prove.

In the longer term, there is a tendency for minimum economic scale in farming to be at a size which can be operated by 1-2 people using modern techniques. Farms below this size tend to be at a competitive disadvantage and may either grow or eventually cease trading—although many continue to exist in favourable marketing niches, or as part of a larger farm. If the number of birds which would be managed by one person were to fall, this could reduce the threshold below which farms tend to feel pressure to grow larger, known as the minimum economic scale.

This does seem to imply that non-cage units would be smaller than batteries. However, a number of points need to be borne in mind when evaluating this argument.

Table 43 shows the size distribution of UK laying flocks. More than a quarter of the laying flock is currently in units of less than 10,000 birds, although unit size is increasing rapidly. Furthermore, as Table 44 indicates, more than a quarter of retail egg sales are made at the farm gate or by a farm roundsman, and this figure has proved remarkably resistant to rising motoring costs. Smaller unit size would not necessarily add to these marketing costs.

Table 43. LAYING FLOCK STRUCTURE

Proportion of the UK laying flock by size of flock

Size of flock	1971	1975	1978
Under 500	11	6	4
500— 999	4	2	1
1000— 4999	22	14	11
5000— 9999	17	14	12
10000—19999	15	17	15
Over 20000	31	47	57
	100	100	100
Total layers ('000)	53855	49360	50489

Source: *Poultry World 17th July 1980*

Table 44. HOUSEHOLD EGG PURCHASES BY OUTLET TYPE (1979)

	%
Retail Shops	50.9
Farm direct/roundsmen	26.2
Milkmen	6.7
Other; including market stall, etc.	16.2
	100.0

Source: Poultry World 17th July 1980

It is important not to exaggerate the decrease in farm size which might occur as a result of changing systems. In the more competitive systems, such as deep-litter and aviaries, a worker can cope with almost as many birds as he could with battery cages. Furthermore, if economies in marketing are important then this would provide an incentive to maintain unit size. Where necessary, units would tend to expand to minimise marketing costs.

It is clearly difficult to predict the effect that changing the system of production will have upon distribution costs. It is most unlikely that changes in distribution costs will be parallel to changes in production costs, and they may well be fairly insignificant.

The following Table is based upon 'guesstimates' of distribution margins, intended to give a more reasonable indication of likely margins than either a percentage 'mark-up' or a fixed margin would give.

Table 45. ESTIMATES OF RETAIL PRICES OF EGGS PRODUCED UNDER DIFFERENT SYSTEMS OF MANAGEMENT

	battery	deep-litter	aviary	straw-yard	semi-intensive	range
production cost	0.45	0.48	0.49	0.54	0.58	0.67
distribution margin	0.25	0.25	0.25	0.27	0.28	0.30
retail price	0.70	0.73	0.74	0.81	0.86	0.97
per cent greater than battery cages		4%	6%	16%	23%	38%

4.2 Retailers survey

In June 1980, we undertook a postal survey of shops selling 'free range' eggs to obtain information on prices and margins. Sixty-six per cent of the questionnaires sent out were returned, a total of 94 useful responses. Retailers, whose names were all obtained from the Free Range Egg Association, were asked how much they charged for eggs and at what price they purchased from suppliers. This information is given in Tables 46 and 47. A large gap in pricing practice between London and the provinces quickly became apparent, and the two are treated separately. On average, retail prices of non-cage eggs were 30 per cent more than cage eggs outside London and 45 per cent in London. However, the

Table 46   RETAIL EGG PRICES—JUNE 1980

		Out of London		London	
Size	Cage	Non-Cage	% Greater than Cage	Non-Cage	% Greater than Cage
1	74p	98p	32	£1.10p	49
2	71p	88p	24	99p	39
3	64p	88p	37	97p	52
4	62p	79p	27	90p	45
5	59p	75p	27	82p	39
		weighted average	30%		45%

Table 47. PRICE OF EGGS TO RETAILER—JUNE 1980

Size	Cage Pre-packed*	Cage Keyes† trays	Non-cage	% Difference Pre-packed*	% Difference Keyes trays†
1	65p	62p	78p	20	26
2	63p	59p	71p	13	20
3	59p	54p	70p	19	30
4	57p	52p	64p	12	23
5	53p	48p	57p	7	19
		weighted average		14%	24%

* packs of half a dozen eggs.
† open trays of 2½ dozen eggs.

price paid by retailers was considerably closer to the cost of cage eggs and there was no significant disparity between regions.

The price of non-cage eggs to the retailer was, on average, around 14 per cent more than the cost of pre-packed cage eggs, and 24 per cent more than cage eggs supplied on Keyes trays*. Non-cage eggs are sold in both forms but unfortunately our survey did not ask retailers to specify how their eggs were sold.

The proportion of the egg price taken as the retailers' margin was 26 per cent in London and 19 per cent in the provinces. This is considerably greater than for cage eggs, where margins are a little below 10 per cent. This implies that an important reason why 'free-range' eggs are more expensive in the shops is because retailers take a high mark-up, twice as much or more, than for cage eggs. This may reflect the way in which they are marketed—in low turnover outlets, often alongside 'health' foods which are notorious for carrying very much higher margins than foods in general. Only rarely are 'free-range' eggs available in supermarkets. Just 5 per cent of respondents in our survey sold more than 150 dozen per week and half sold under 50 dozen.

Unlike battery eggs, there is often only one retailer of 'free-range' eggs in a locality, indicating a lack of competition. Furthermore, supplies can be difficult to obtain. Retailers were asked whether they could obtain sufficient eggs to meet the demand. The position varied: 38 had no difficulty; 41 could usually obtain supplies; 5 had difficulty during winter; and 4 could not obtain enough eggs at any time.

It is interesting to compare the results of our theoretical costings developed in section 2 of this chapter, with the information in the retailers survey. Eggs sold as 'free-range' under the rules of the Free Range Egg Association are not allowed to come from deep-litter or

*i.e. open-trays containing 2½ dozen eggs

aviary units and are supplied largely from semi-intensive systems. The actual price of non-cage eggs to the retailer is in the range 14-24 per cent greater than cage eggs. Our theoretical costings place the cost of semi-intensively produced eggs around the top of this range, and are therefore consistent with the survey results.

4.3 Conclusion

The figures presented in this section suggest that there is very little difference between the retail price of battery eggs and what one would expect from the intensive flock systems—deep-litter and aviary. The less-intensive flock systems would increase retail prices more significantly: straw-yard by about 16 per cent, semi-intensive by about 23 per cent and free-range by about 38 per cent. However, the comparison is sensitive to assumptions about the number of birds which are kept in a battery cage. If this were limited to fewer than five in a standard cage, this would significantly alter the comparison in favour of the less-intensive systems.

In the light of these figures it is at first sight surprising that 'free-range' eggs are often so expensive in the shops. The Earth Resources Research survey of 'free-range' egg retailers suggested that this could be explained by the high mark-up generally applied to 'free-range' eggs at the present time as a result of them being marketed as low turnover 'luxury' goods.

4 Some wider costs — energy and livestock wastes

Although this book focusses upon the 'on-farm' implications of different methods of livestock husbandry, it is important to be aware that the economist's interest is wider than this. As well as the direct costs to the farmer of such items as labour and feedstuffs, there are also indirect or 'external' costs which are imposed upon society as a whole. Most obviously, intensive livestock production has been identified as a source of environmental costs. Units are associated with noise, smell, flies, rodents and heavy traffic, generating a more acute form of nuisance than more traditional farms, partly because of the techniques involved and partly because animals are kept in larger numbers and in smaller areas than before. In this chapter the problems associated with handling large concentrations of livestock wastes in liquid form are considered.

Energy costs are a second area where there may be significant externalities. It is commonly felt that the 'true' costs of depleting our energy resources are not reflected in their price. In the second part of this chapter we try to establish whether energy costs may provide a significant basis for deciding between alternative systems of livestock production.

1. Livestock wastes

Livestock produce very large quantities of effluent. 1000 pigs or 20,000 hens pose a disposal or treatment burden equivalent to that of a town of 2000-2500 people. In the past this was not a problem; manures were coveted as a fertilizer and provided an important justification for keeping animals. However, the introduction of artificial fertilisers during the Second World War greatly reduced the dependence of arable fertility upon livestock, at least in the short term. Subsequently there has been a strong trend towards specialisation so that, regionally, the most productive crop lands produce arable crops continuously and livestock is kept on land less suited to arable use. Individual farmers have tended to focus on a single or small groups of enterprises to take advantage of economies of scale.

Three specific developments are important here: livestock units have grown much larger; their links with surrounding farm lands are becoming more tenuous; and wastes are increasingly handled in liquid rather than solid form, as slurry rather than manure.

As a result of these changes, waste disposal is becoming a serious problem, especially to the pig and poultry sectors. Slurry is a particular problem, and is often associated with the more intensive forms of livestock production. It is normally applied to the land without any kind of treatment.

Of more than 400 pig farms handling slurry in a recent survey undertaken by the NFU, only 10 per cent carried out any treatment before dispersal.[1]

Because slurry is liquid it is particularly likely to lead to surface run-off into water-courses, which can cause eutrophication and a build up of nitrates. The problems of surface run-off are particularly severe where application rates are excessive or where slurry is applied to frozen, water-logged or sloping land. The mechanisms causing and implications of eutrophication and nitrate build up are not clearly understood, but both cause concern. Livestock wastes have almost certainly contributed to the eutrophication which is causing serious ecological damage in the Norfolk Broads, as well as other inland waterways.

Concern over high nitrate levels in water supplies stems from the risk of causing methaemoglobinaemia in bottle-fed infants— the 'blue baby syndrome'— and the possibility that ingestion of nitrates will result in the formation of carcinogens in the body. Evidence is not conclusive in either case. However, some water authorities face difficulties in consistently supplying water with nitrate levels at or below those recommended by the WHO, and nitrate levels in aquifers and rivers appear to be rising. Therefore changes in agricultural practice which contribute to higher nitrate levels are likely to continue to attract attention.

Further pollution problems are caused by the presence of disease organisms in slurry. Untreated slurry breaks down anaerobically*, reaching rather lower temperatures than under aerobic conditions†. Consequently disease pathogens, Salmonellae being the most notable, are capable of surviving for much longer in it—in some cases approaching a year.[2]

A survey of slurry from 54 pig farms found more than a fifth of farms

---

* in the absence of oxygen
† in the presence of oxygen

to be infected with Salmonella.[3]

Where run-off occurs, these organisms can contaminate water supplies. Heavy applications of slurry have been shown to give rise to increases in the presence of faecal organisms in running water within 2 hours of spreading.[4]

Equipment used for spreading which projects slurry into the air can lead to organisms drifting considerable distances downwind.

Smell is the most noticeable pollution problem from slurry. Anaerobic fermentation gives rise to a number of highly odorous gases which are released when the slurry is distributed, and particularly when it is spread. The NFU admitted in a submission to the Royal Commission on Environmental Pollution that 'problems of odour pose one of the most difficult areas of control and are one of the most common sources of friction between the agricultural industry and other rural activities'.[5]

Hodge, in a survey of livestock producers in Surrey, found smell to be the most frequent source of complaint.[6] Of 2452 'odour incidents' reported in a survey undertaken by MAFF and the District Councils Association, over 70 per cent were related to livestock buildings and land applications, with pig units being the most commonly cited.[7]

Quite apart from the wider environmental aspects, heavy slurry application can have a detrimental impact upon agricultural activity. It can lead to compaction, poaching and blocking of pore spaces in the soil. This creates anaerobic soil conditions, inhibiting root growth and germination. Heavy applications result in smothering and tainting of herbage which reduce palatability for grazing livestock. In addition it can result in excessive nitrogen uptake by plants. In cereals this leads to 'lodging', where excessively lush growth of the foliage overburdens the stem making the crop vulnerable to wind and rain. Finally, it has been observed that high concentrations of hydrogen sulphide, ammonia and carbon dioxide in livestock buildings, given off from under-floor slurry stores, can depress performance and in extreme cases lead to the death of both stock and stockman.

Many of these problems can be overcome by good practice and the use of available technology. Various means of treating slurry exist to render it safer and less offensive. One way is to mechanically separate it into solid and liquid fractions. The solid fraction is relatively odour free and can be stacked and stored for long periods without difficulty, while the liquid can be pumped directly onto the land via irrigation pipes whilst still fresh and comparatively odourless. More satisfactory methods are to aerate the slurry or mix it with straw to compost. These ensure a predominantly aerobic breakdown, and the high temperatures generated destroy pathogens. When spreading on the land, care can be taken to ensure that the application is not too heavy, and that the slurry is directed close to the ground. If slurry is applied at the most appropriate time of year, in accord with crop nutrient needs, this will ensure that as much as possible is taken directly for crop growth and therefore is not available to leach or run from the soil. In the case of grassland this means spreading early in the spring and again immediately after silage or hay-making. In order to have this flexibility in management, it is necessary to have the facilities for long-term slurry storage available. Stores must be well constructed and ideally protected from rainwater—leaking and overflowing slurry stores are a major cause of pollution and complaint.

In practice, although these means of reducing the environmental impact of livestock waste exist, they are not widely adopted. As the NFU admitted to the Royal Commission on Environmental Pollution, only a very small fraction of livestock waste is in fact treated or processed in any manner.[8] As treatment can be a very costly process, it is not surprising that farmers have been less than enthusiastic. A guide to comparative costs of waste handling systems for fattening pigs was provided by Nielson (1980) based on work underway at the National Institute of Agricultural Engineering.[9]

Table 43. ESTIMATES OF WASTE HANDLING COSTS FOR PIGS (1980)

Waste handling system	Cost per pig place p.a. (£)
Slurry-tanker direct from building to land weekly	2.09
Slurry from building to 6 month tanks then to land	4.97
Danish buildings, strawbed 6 months store to land	11.09
Flush cleaning, anaerobic treatment, solid wastes 6 months storage	10.00

These figures are rather sensitive to assumptions about depreciation rates, interest charges, utilisation levels and spreading distances, and so should not be taken too literally. However, they provide a useful illustration of the extra cost of treating wastes so as to limit their impact upon the surrounding environment.

Some legal and planning mechanisms exist by which the activities of livestock farmers can be constrained in order to minimise pollution. There is legislation to control the discharge of effluents to watercourses, and an action against smell nuisance can be taken under Public Health legislation. This provides a means of taking action once pollution has occurred. Preventative action can be taken in some places by planning authorities faced with new intensive livestock enterprises by refusing permission or attaching conditions to their development.

Humberside County Council, faced with a very dense and rising livestock (mainly pig) population, is a notable example of an authority which has recently incorporated an Intensive Livestock Units Subject Plan into its Structure Plan.[10] The Plan includes stipulations that: intensive livestock units should not be established close to residential dwellings; sufficient suitable land should be available to take the slurry; and sufficient storage should be provided so that slurry need only be spread when conditions are appropriate. However, this procedure only affects new development. Moreover, most buildings of less than 465 square metres do not require planning permission anyway.

In reporting on Agriculture and Pollution, the Royal Commission argued that intensive livestock units should be regarded as industrial enterprises, and must come to accept the cost of pollution control in the same way as any other industry. External cost should, in other words, be internalised. Their recommendations included amendments of the General Development Order to exclude intensive livestock units, whatever their size, from exemption from planning permission.

Intensive livestock farmers may well be forced to respond to increasing public pressure to control the nuisance and pollution which their unrestrained activity creates. This means either more sophisticated treatment of slurry or a resurgence of interest in traditional methods of waste handling and use. This will greatly reduce the cost advantage of existing slurry systems, to the comparative advantage of those less-intensive systems which do not handle waste in this form.

2. Energy use

Intensive livestock units are often regarded as very energy intensive. They substitute mechanical power for human labour and provide a closely controlled environment for the stock. It is sometimes argued that such systems, developed in an era of comparatively cheap energy, may be vulnerable now that energy prices are rising in real terms. In this section we examine whether intensive livestock systems are, in fact, especially profligate in their use of energy, compared to less intensive forms of husbandry.

If we examine the amount of energy which is put into the farming system, both directly in the form of fuel and indirectly as machinery, buildings, fertilisers etc., then livestock production does require a lot of this support energy in relation to the amount of protein or food energy which it produces. As Table 49 shows, livestock products require a great deal more support energy per megajoule (MJ) of food energy or

Table 49. ESTIMATES OF AGRICULTURAL USE OF SUPPORT ENERGY

	MJ of energy required to produce:	
	1 MJ of food energy	1 kg of protein
wheat	0.3	45
potatoes	0.7	113
milk	1.4	118
pigs (pork and bacon)	1.6	238
eggs	3.8	200
broilers	6.7	203

Source: White (1975)[11]

kilogram (kg) of protein produced than, say, wheat or potatoes.

However, although livestock production absorbs a lot of energy, it uses very little directly. The inefficiency of livestock production in energy terms is mainly due to the conversion losses in the use of feedstuffs which themselves require large support energy inputs to produce. Three-quarters of the primary energy consumed in UK agriculture is accounted for by fertiliser, petroleum and feedstuff processing. Much of this is used in growing and processing livestock feed before it ever reaches the livestock unit. The non-feed support energy actually used on the livestock farm to provide light, heat and power, and embodied in the machinery and buildings, is only a small part of the total energy used to produce meat, milk and eggs.

This has two implications for our comparison between livestock systems. Firstly, differences in non-feed-related energy use would have to be very large to greatly affect the difference in overall energy use. Secondly, differences in feed efficiency will be critical in determining the relative energy efficiency of different systems. Furthermore, it may be that where energy is used directly in ways which will improve feed efficiency, for example by providing heating or more elaborate buildings, overall energy efficiency on the farm could actually be improved. The relationship between system and energy efficiency is not a simple one.

It is likely that the energy needs of intensive systems have been exaggerated. Take the case of energy used in the manufacture and construction of buildings. As a simplification this is often assumed to be proportional to capital costs. With the exception of outdoor pigs, less-intensive systems tend to have similar or greater capital costs than intensive methods. Thus, the energy costs of the fixed equipment per animal may be about the same, although in practice it is unlikely that energy costs really are proportional to capital costs in this way. It seems likely that the cages and stalls of intensive systems have a comparatively high energy 'density', although details are not available.

It is still unusual for energy to be used to heat livestock buildings, except for very young and vulnerable stock, although the increasing use of early weaning for pigs is making this a more significant energy use. In general, intensive systems maintain temperatures by good insulation and dense stocking to conserve the animals' own heat.

Lighting is normally provided for all egg production systems and most indoor methods of pig keeping. In addition, controlled environment systems require some electricity to power fans for mechanical ventilation, although this is small. Fan ventilation accounts for about 8.5 per cent of total energy used in battery egg production.[12] Total electricity used in broiler production is just 4.6 per cent of the entire energy input.[13]

Extensive pig and poultry systems require feed to be taken greater distances to stock, requiring tractor fuel. Semi-intensive systems may also use comparatively large amounts of fuel for handling wastes as manure.

On balance, it is unlikely that energy efficiency is a significant reason for choosing between systems which are in commercial use today. This is borne out by a detailed analysis of alternative systems of egg production carried out by Wathes (1981), some of whose results are tabulated in Table 50.

Table 50. POSSIBLE FUEL ENERGY INPUTS PER BIRD*

	Battery		Strawyard		Free-range	
	MJ	% of total	MJ	% of total	MJ	% of total
Feed	449.2	71	489.1	77	505.3	75
Electricity & fuel	132.4	21	67.8	11	88.5	13
Buildings	30.2	4.5	45.6	7	47.0	7
Miscellaneous	22.3	3.5	29.0	5	36.2	5
	634.1		631.5		677.0	

* including energy inputs for pullet rearing
Source: Wathes (1981)[14]

5 Changing systems

In this book so far a range of intensive and less-intensive systems have been compared. While the costings have illustrated that the extra 'cost' of some less intensive systems is a great deal less than is often supposed, it is clear that in the absence of changes in the law or the structure of grants there is unlikely to be any large scale abandonment of intensive systems in the near future. While outdoor pig production may become more popular as a means of avoiding high capital costs and there is likely to be an expanding market for free-range eggs, the central position of intensive systems is unlikely to be altered by market forces alone.

A number of factors re-inforce the strength of the position held by intensive systems, and they are considered briefly here.

Consumer choice. The consumer needs to be well informed in order to choose effectively between the products of different systems. However, it is usually very difficult for consumers to know how the livestock products on sale to them have been produced. Even in the case of 'free range' eggs the opportunities to abuse the label are obvious. The differences between systems are often deliberately blurred when livestock products are marketed. They are given a 'traditional', 'country goodness' image whatever their source.

Producers margins. Pig and poultry production have become increasingly competitive and the profit margin per animal has become very small. Consequently, small changes in costs or performance, which might affect prices to the consumer only slightly, will have a more substantial impact upon producers' profits. Even where farmers would prefer one system over another, perhaps for welfare reasons, the narrowness of the profit margins within which they are working does not allow a great deal of choice.

Labour requirements. In general, the labour requirement is often greater for less-intensive systems. Where good, responsible and skilled stockmen are difficult to find and keep, the farmer who wishes to minimise his reliance upon non-family labour is likely to be attracted to systems which substitute capital for labour. Furthermore, it can be argued that less-intensive systems present a greater challenge to stockmanship. There are fewer mechanical aids, less environmental control, and fewer constraints on the behaviour of livestock. Perhaps modern farmers are less interested in the role of stockman and would rather concentrate on the challenges posed to the technician and the businessman.

Tax reliefs and grants. It has often been claimed that tax-reliefs and grants for capital expenditures have favoured intensification. However, this claim has little foundation. In the first place, as we have seen, less-intensive systems often require similar amounts of capital investment to intensive systems and so stand to benefit equally from reliefs and grants. Secondly, the poultry industry no longer receives any capital grants. Grants to the pig industry are restricted to farms capable of supplying at least 35 per cent of their feed, and only the smaller pig herds are now eligible. And finally, while tax reliefs on capital expenditures undoubtedly assist investment, this is comparable to setting other business expenditures against income, and these allowances are available to industry as a whole. It is difficult, therefore, to regard them as biased in favour of intensive livestock production.

However, the dominant position of intensive systems is not unassaillable. The possibility of introducing farm animal welfare legislation of a kind likely to alter the balance between systems is now being widely debated. Discussion is taking place in the EEC about regulating cage space for layers which, as was shown in chapter 11, can significantly alter the attractiveness of cage systems compared to the alternatives. The Agriculture Select Committee of the House of Commons reporting in July 1981[1] advocated legislative changes to outlaw certain practices and systems, although these have not been taken up by the Government.

It is appropriate, therefore, to give some consideration to the implications of changing from intensive to less-intensive systems on a wide scale. For example, it is sometimes argued that straw and land are too scarce to allow a widespread change to extensive systems. These arguments are examined first. Resistance to change is to be expected from some producers and consumers. The interests of these two groups are considered. Finally, there would be once and for-all costs associated with a transition compelled by legislation. Although it is not possible to estimate these accurately, their likely significance is discussed.

1. The use of land

At a time when the opportunity cost of land is high, should it be used for ranging animals which could be kept efficiently indoors?

If the entire laying flock of 50 million birds were to be kept under semi-intensive conditions, with outdoor runs stocked at 2000 birds per acre, this would require an extra 25,000 acres to be made available for the exclusive use of the UK poultry industry. This is equivalent to less than one year's loss to urban development at the present rate.[2] Under free-range, at 150 birds per acre, the entire laying flock would require access to one third of a million acres, a little more than 1 per cent of the area under crops and grass in the UK. However, under free-range conditions the opportunity cost is less clear-cut because there is a shared land-use. Free-range hens, stocked at 150 to 250 birds per acre clearly would not fully use the capacity of grassland to support livestock. Other stock would need to be grazed alongside or in rotation. Indeed, George Henderson writing in the 1940s was very enthusiastic about the positive value of ranging chickens:

'So valuable are poultry for reclamation, in scratching out moss, treading down rough broom grass and leaving their rich manure behind, that on this poor light land it is now possible to carry a beast to the acre, grazing before and after the pens, where three acres was insufficient and had to be supplemented. . . So highly do we value poultry for fertilising purposes that we would consider their retention justified if they did not leave a profit in their produce.[3]

Other sources of fertility are now more readily available to substitute for the work of the chickens, but the point remains that there may be little conflict between low-density free-range birds and other land-uses.

Both pigs and poultry may be used to some extent as a catch crop, cleaning up after a cereal or forage crop before the land is prepared for another crop. Where this can be worked into a rotation without displacing other uses, the land required to range the livestock is effectively 'free'.

Outdoor pigs are often used as a break crop on land which is unable to sustain continuous cereal growing. In these circumstances the opportunity cost is not a high value cereal crop, but perhaps a root or grassland break crop. Pigs are an attractive break crop on light soils which tend to be deficient in the organic matter which pig waste provides. It is widely agreed that subsequent crop yields can be significantly improved, although quantification is difficult. Boddington, who was particularly interested in the use of pigs as a break crop, was unable to obtain useful quantitative data from the farmers in his survey; a number considered that there were significant yield benefits, although there were a few who felt that subsequent yields might even be harmed.[4]

More recently, one large pig producer who rents land on the Breckland Sands has been able to rent at £30 per acre compared to the £80 asked for intensive carrot growing.[5] But this is just one soil type, and it is not clear how far the same benefits could be achieved simply by spreading the wastes from an intensive unit on the land. Indeed, one disadvantage of ranging pigs is that they tend to spread their muck unevenly. On the other hand, they charge nothing for spreading it!

A further point is that the livestock can derive some nutrients from the land. This is not easy to quantify and is clearly going to vary considerably according to crop and season. The costings presented earlier show that free-range hens have a lower feed consumption per bird than under semi-intensive and straw-yard conditions and only a little more than in cages. This is partly due to the contribution from the pasture in the form of insects, seeds and a little grass, particularly during the warmer months of the year. However, this is more than offset by the lower yield on free-range, so that feed per egg produced is highest outdoors, some 20 per cent more than in batteries. Sows kept at pasture are able to derive nutritive value from the grazing. This is potentially quite considerable during the growning season, amounting to 25-50 per cent of nutrient requirements.[6] However, in practice farmers often do not manage to achieve these savings.[7] Further, in winter, where conditions are difficult and the value of forage is reduced, the extra energy required to maintain body temperature exceeds the nutritive benefit gleaned from the forage. Group feeding, which is generally practised outdoors, tends to mean that feed requirements are higher if weaker sows are not to lose condition. On balance, survey data suggests that feed consumption may be slightly higher outdoors, but that this may be offset by higher piglet weights.

So, although ranging outdoors does mean that the land is providing food for the pigs or hens, this cannot normally be counted as a benefit as the requirement for concentrated feedstuffs is no less, and sometimes more, than indoors. There is one possible exception to this. Pigs could graze crops such as roots, which have an output of both energy and protein per acre which is considerably higher than for cereals. While they can only be partially substituted for cereals in the pig's diet, because of their low energy density and the limited capacity of the pig's gut to cope with bulky foods, there would be some scope for improving the productivity of land by grazing pigs on roots rather than growing cereals to feed the pigs. However, this is not a system which is used commercially.

From this discussion we can conclude that a simple comparison between the land requirements of different production systems would be misleading. However, it is evident that some systems are compatable with other land uses and even when this is not the case, the additional land needed is not that great.

2. The availability of straw

Farmers who adopt systems which do not require straw sometimes do so to avoid dependence upon a raw material which can be costly and difficult to obtain in sufficient and regular quantities. It is sometimes difficult for an observer to reconcile this with the clouds of smoke darkening our late summer horizons as straw is burned in the field.

This section examines the amount of straw produced in the UK and how much of it is burnt on farms. The reasons for strawburning are considered to see whether there are strong agricultural arguments for this practice or whether it is simply a means of disposing of surplus straw.

2.1 The production of straw in the UK

Accurate figures for the amount of straw produced in the UK do not exist. Production varies not only according to cereal acreage and yield, but also sowing times, varieties grown and length of stubble. The estimates in Table 51 were given at an ADAS conference in 1980.

Table 51. TOTAL STRAW PRODUCED IN THE UK (1978)

	Yield (tonnes/ha)	Area (million ha)	Total Yield (million tonnes)
Barley	2.8	2.36	6.61
Wheat	3.8	1.26	4.70
Oats	2.8	0.18	0.50
TOTAL		3.80	11.81

Source: Power Farming, April 1980

The straw yields achieved in 1978 are close to what appear to be average levels, although in a bad year they might be 30 per cent lower.[10]

Much of this straw already finds a use in livestock husbandry. It is likely that some 45 per cent of dairy farms still use straw bedding in the traditional way, and surveys of pig farms showed 35 per cent using a straw-based system, and a further 40 per cent using a combination of straw and slurry.

Barley and oat straw are used as a feed for ruminants and there is a developing market for wheat straw which has been nutritionally enhanced by treatment with an alkali. There is also a growing, if small, market for straw in non-agricultural uses— strawboard for example— and straw has been identified as a raw material for paper-making and possibly as an energy source. Nonetheless, substantial quantities never leave the field and are treated as a waste product to be burned or ploughed into the ground. Some idea of the amount of straw going into different uses was obtained when ADAS carried out surveys of 430 farms in 1976 and 524 in 1977 to determine the fate of cereal straw in England and Wales.[13] Their results, together with estimates for 1972 made by the Advisory Council for Agriculture & Horticulture[14] are presented in Table 52.

Table 52. THE FATE OF CEREAL STRAW IN ENGLAND & WALES

Percentage which is:	1972	1976	1977
Burned	37	21	41
Ploughed or cultivated	2	6	2
Baled	61	73	57

Variations in results can probably be explained to a large extent by weather conditions. 1976 was a year of drought, whereas the post-harvest weather in 1977 was not ideal for straw-handling and hence the low proportion baled. The surveys suggest that 25-45 per cent of cereal straw in England and Wales is burned or ploughed in. This amounts to between 2½ and 5½ million tonnes of straw, depending upon the yield.

2.2 Is there surplus straw?

Many farmers regard straw burning not only as a means of disposing of a waste product, but as a useful agricultural practice in its own right with a beneficial effect upon subsequent crop yields. It is claimed that burning destroys pest and weed seeds and is preferable to ploughing straw residues into the soil. Incorporating straw in the soil can initially lead to a net loss of available nitrogen as this is 'tied up' in the process of decay, only later becoming available as the organic matter is broken down. This is not, however, a problem where nitrogen fertiliser is applied at fairly high levels. Furthermore, if straw residues are ploughed in over a number of years, any dampening effect upon yields disappears as the nitrogen claimed by fresh straw is balanced by that released from the decay of older residues.[15] It is likely, therefore, that the deleterious impact of ploughing-in straw residues upon the nitrogen available in the soil has been exaggerated, particularly in the case of farms where it is a regular practice.

Straw-burning is an effective way of reducing the number of weed seeds on, or just below, the soil surface, perhaps to an extent which would justify reducing herbicide applications. However, seeds on the surface are only a small fraction of the total numbers within the soil capable of germination.[16] The effectiveness of burning in reducing the innoculum and host material of certain diseases has not been similarly established.[17] Recent research at Rothamsted Experimental Station has failed to find that straw-burning has any overwhelming harmful or beneficial impact upon soil fauna.[18] While populations of pests, such as cereal aphids, are reduced, numbers of certain predators, spiders for example, are reduced also. Earth worm populations, too, can be greatly reduced by continuous burning.

There have been some attempts to measure the effects of different methods of straw disposal upon subsequent crop yields in trials at the Experimental Husbandry farms. Staniforth summarises their results as follows: "the yields of winter cereals grown immediately after cereal crops can benefit considerably from straw and stubble burning at least when they are direct-drilled or sown after minimal cultivation. For spring-sown cereals, or for cereals sown in land which has been ploughed, there is little or no evidence of benefit from burning, although it will always be necessary in the short-term to take account of the nitrogen factor when straw is incorporated into the soil".[19] It is, therefore, minimally cultivated direct-drilled cereals which have been identified as most likely to benefit from straw-burning. However, this need not mean that under these conditions straw is best not harvested. Providing that weather conditions are suitable, a *stubble* burn can be as effective as a straw-burn in getting rid of trash in preparation for the next crop.[20]

While farmers are conscious that straw-burning may have some advantages for crop yields, burning takes place largely because there is insufficient financial incentive to justify harvesting straw. Baling is more costly than burning, and more likely to interfere with the cultivations needed for the next crop. Moreover, it requires a lot of labour at what is already a peak time, although the development of more mechanised bale handling systems has helped to overcome this problem. Transport and storage impose additional costs, although these might be borne by a contractor or purchaser.

It is significant that, as Table 53 shows, the proportion of straw baled is higher for barley and oats than for wheat. This reflects the value of the former as feed for cattle, and its relative scarcity in regions where the cereal acreage is lower in relation to the livestock populations. In other words, the amount baled reflects its scarcity quite closely, which suggests that it is fairly responsive to price. This is borne out by a survey of farms undertaken in Eastern England by Morris et al (1976).[22]

Table 53. AMOUNT OF STRAW BALED BY REGION AND CROP (1976)

(% of total cereal area)

	wheat	barley	oats
Northern	61	96	99
Yorks/Lancs	38	86	89
Lincs	28	87	100
W. Midland	80	94	99
Eastern	14	84	98
S. Eastern	79	98	98

Source: ADAS (1976)[21]

Interpretation of the supply curves which they derived from their results suggests that, at 1980 prices, it would be necessary to offer farmers ex store straw prices of approximately £17 per tonne to obtain 50 per cent of the straw presently burned, and £20 per tonne for 90 per cent. These figures ought not to be taken too literally as they rest upon some arbitrary assumptions and a small sample. However they do suggest that were more straw to be required by livestock farmers, there would be a fairly elastic supply response.

To conclude, it appears that not only are there large amounts of un-utilised straw at present, but also that farmers would be willing to make much of this available in response to a modest increase in price.

3. The producers' interest

It is often supposed that the interests of farmers and of the animal welfare lobby are in conflict because the restraints proposed by the latter would inhibit farmers in running their businesses profitably. To remain competitive, it is argued, farmers must not be prevented from innovating and adopting the most profitable techniques.

While it is true that farmers need to remain competitive, it is doubtful that they themselves benefit from this process, other than in the negative sense of remaining in business. In a competitive market, falling production costs will tend to be passed on to the consumer. The most progressive farmers will enjoy enhanced profits for a while as a result of introducing lower cost methods but, as more farmers adopt new techniques, production costs in the industry generally will fall. Lower costs encourage an expansion of output so as to take advantage of the apparently enhanced opportunities for profit, but prices will respond by falling as supply expands. Over time, this process will tend to leave farmers' incomes at a fairly constant level.

Do farmers benefit from this process? Those farmers in the vanguard of innovation will tend to earn rather better profits as a result. However, there will be another group slower to innovate whose incomes will suffer accordingly. Overall, the lower prices which less costly production methods allow will stimulate demand for the industry's products, enchancing incomes and employment in the industry as a whole. However, this is likely to be more than offset by the saving in labour which is often associated with new techniques and which underlies the increase in enterprise size and the corresponding fall in the total number of enterprises.

Of course, discouraging the use of certain intensive methods would not end this process but rather readjust its focus. If less-intensive techniques were more widely used, this would shift the attention of innovators to developing the potential of these systems. Furthermore innovation associated with the adoption of intensive techniques has had much less effect on performance than have more widely applicable advances such as improvements in breeds and in feed formulation.

The producers whose interests would be most damaged by discouragement of certain intensive practices are those who have invested in equipment, training, or methods of organisation which are unique to those practices; and farmers unable to readily adopt new systems because of such contraints as insufficient land or problems with obtaining planning permission.

There is one important qualification. If some intensive techniques were to be discouraged in the UK but not elsewhere in Europe, this could put UK producers at a competitive disadvantage.

However, one should not exaggerate the extent to which farmers within the EEC are subject to common competitive pressures; examples of unilateral discrimination abound in the form of subsidies, grants and health regulations. Nonetheless, international competition, certainly within Europe, is an important factor. There would therefore be advantages in seeking a common EEC approach towards the regulation of intensive systems.

4. The consumers' interest

Price is undoubtedly one of the most important influences on consumer purchases of animal products, especially meat. However, having said this, it must be stressed that people are generally reluctant to switch from meat to alternative high protein foods unless there is a considerable change in the price differential between the two. On the whole, consumers are more likely to respond to price changes by switching between kinds, cuts and qualities of meat than from meat to other types of food. In the case of eggs, people are even less likely to react to increased prices by buying different foods. The demand for eggs is fairly insensitive to price changes and can be described as "inelastic".

A second major influence on demand for meat and poultry products is changes in income. As incomes have risen, so has the demand for meat. This has not been the case with eggs, reflecting their low income elasticity of demand. As well as increasing the amounts purchased, consumers have opted for more expensive types and cuts of meat— steaks rather than mince, lamb in place of mutton. Indeed, it seems reasonable to suppose that if incomes rise the emphasis on quality will become more important as quantity is clearly limited by appetite.

Many attributes of meat contribute to people's perception of quality. Taste, tenderness, aroma and freshness are all regularly named. Whether rightly or wrongly, consumers associate quality to some extent with the system under which animal products are produced, perhaps in a rather vague way. Furthermore, there is direct concern amongst some consumers for the welfare of the animals themselves, which can influence their purchases. It may be that this will also assume greater significance if incomes rise in the future.

There is not a lot of specific evidence about the significance of livestock management systems for consumer behaviour. Producers seem to be aware of it nonetheless. It seems fair to presume that the images promoted in marketing eggs and broiler meat, of plump, contented, straw-chewing farmers and chickens, reflect a supposition that this is significantly more acceptable to consumers than the rather more prosaic reality. Images which are 'traditional', 'natural', 'farm-fresh' and sometimes 'free-range', are deliberately created and projected by those marketing meat and eggs, with varying degrees of accuracy! The 'white' veal market in the UK probably suffered most from consumer resistance to management practices. Volac, through their subsidiary Quantock veal, set out to develop for commercial use a straw-yard system which would be more acceptable. They regard welfare aspects as an important part of their marketing strategy, although their system also compares favourably in terms of profit to single-housed systems.

In September 1980, Social Surveys (Gallup Poll) Ltd undertook, at the request of the Farm Animal Care Trust and the Universities Federation for Animal Welfare, an independent survey of consumers. A sample of 937, representative of the female population of Great Britain aged 16 and over were asked:

'Would you or would you not be willing to pay more for eggs which you knew were produced in a non-battery cage system? If yes— how much more would you be willing to pay for a dozen eggs?'

The results are summarised in Table 54.

Table 54. WILLINGNESS TO PAY EXTRA FOR NON-BATTERY EGGS

		percentage of sample
Prepared to pay more: total		63
1-5p a dozen more	21	
6-10p a dozen more	24	
11-15p a dozen more	4	
over 15p a dozen more	9	
don't know how much more	5	
Not prepared to pay more		29
Don't know		7

Less than a third of the consumers sampled claimed to be unwilling to pay more for non-battery eggs, and more than 40 per cent reported that they would be willing to pay in excess of 5 pence per dozen more. Of course, not everyone reacts in the same way at the supermarket shelves as they do to an opinion poll interviewer. Nonetheless, the survey does provide some guide to consumer attitudes, and suggests that a willingness to pay more for produce from more acceptable systems is quite widespread.

5. Transition costs

As well as influencing the cost of production, changing from one system to another will impose transition costs. Some items of capital equipment, such as battery cages and sow stalls, are unsuited to other systems and would have to be scrapped. Farm layouts planned for one system might not be as well suited to other methods, imposing a management handicap. In addition accumulated skills and experience relating only to redundant systems would become valueless. The problem is not that new investments would have to be made—these would only occur if they could be commercially justified—but that some existing investments might have to be written off prematurely.

There are three main factors governing the importance of transition costs:

(a) *The amount of capital that becomes obsolete.* This clearly depends upon the extent to which intensive systems are discouraged. Quite a lot of intensive buildings and equipment could be converted to less-intensive use, although in some cases conversion may not provide the conditions best suited to achieving good results.

(b) *The remaining economic life of the capital,* had it not been rendered prematurely obsolete by changed welfare regulations. This might not be very long for battery cages; much of the existing stock was installed 10 or more years ago and is reaching the end of its life. Uncertainty over the future acceptability of cages has probably deterred re-investment. The position is likely to be different in the pig industry, in which the more intensive methods are still being newly adopted.

(c) *The transition time.* If certain techniques were prohibited, the longer the transition time allowed for changeover, the smaller the transition costs. As farmers will not invest in methods which are to be prohibited, there will be a gradual re-investment in new techniques. This would mean both that the changeover could be fairly smooth, minimising disruption, and that the outstanding economic life of capital rendered prematurely obsolete would be shorter. It would therefore be desirable for there to be a substantial transition time, perhaps 5-10 years, for major changes in welfare regulations, in order to reduce changeover costs.

6. Changing systems in perspective

Various implications of a widespread change to less intensive systems have now been considered. And it appears that these need not constitute major obstacles to change. Although less intensive systems require larger land areas for stock than do intensive systems, the additional requirements are comparatively small and could be accommodated without major changes in land use. Similarly, the availability of straw for bedding would be unlikely to prove a problem. At present farmers dispose of large quantities of straw to little advantage and it appears that fairly small increases in price would encourage farmers to make available much of the straw which is presently burned.

Although producers' organisations are strongly opposed to welfare regulations, producers would not necessarily be damaged by regulations prohibiting certain systems. Provided that regulations did not put UK producers at a competitive disadvantage—and either a common EEC approach or protective measures for UK farmers could ensure this—farmers' profits need not suffer. Some individual producers would be disadvantaged, especially those with a recent investment in techniques subsequently prohibited, but compensation provisions and a reasonable transition period could greatly alleviate this.

On the whole, consumers can be expected to resist increases in the price of food. However they are increasingly becoming concerned about the quality of their food, which they relate to some extent to the system of husbandry. There is survey evidence to suggest that consumers might be willing to pay extra for food which they considered to be more humanely produced.

Finally, the one-off costs of transition from one system to another were considered. These amount to the premature obsolescence of skills, experience and equipment. No attempt was made to quantify these and it would be very difficult to do so. However, allowing an adequate transition period would do much to reduce these costs.

6. Conclusions

Farm animal welfare has become a highly controversial issue. Public concern has been aroused particularly by the more confined and less 'natural' environments introduced for livestock. The industry, while not admitting any deterioration in animal welfare, invokes improved use of resources to justify its practices and strongly opposes any interference in its activities. Economics may not be the most important issue in the debate on farm animal welfare but it certainly has a very important bearing upon government policy and the attitude of the industry and consumers. It is often held that we cannot 'afford' to stop 'progress', and assumed that this maxim provides clear and incontrovertible support for the intensification of livestock production.

While controversy has stimulated some research into the welfare implications of various methods of husbandry, there has been virtually no detailed study of economic aspects. The present study was undertaken to provide some carefully supported comparisons of the economic performance of different systems. Information was drawn from a great many sources, inside and outside the UK, including reports of surveys and experimental work, farm visits, and our own survey of egg producers. These sources were used to develop, as the main focus of the study, theoretical costings of a range of systems of different 'intensity' for comparison. The costings are designed to illustrate the performance of well-managed new investment in commercial-scale enterprises.

This section will briefly summarize the findings of this book, ending with an assessment of their role in the farm animal welfare debate.

1. Pigs

The intensity of production is not a concept which is clearly defined, particularly in the case of pigs, where few systems are in practice identical. There is often a mixture of 'intensive' and 'less intensive' elements on a single enterprise and there are almost as many different systems as there are farms. In weaner production intensity is associated with the degree of confinement, the extent to which straw is provided, and the age at which pigs are weaned. Less-intensive systems use straw bedding, wean at 5-8 weeks, and minimise confinement in crates or cages. Pigs may be kept outdoors for part or all of the year. In contrast, a typical highly intensive system involves weaning young pigs at 2-3 weeks into flat decks, and confining dry sows in stalls. A slatted floor slurry system is often used throughout in place of straw bedding.

The major costs of pig production are feed, capital and labour, and these were examined to see how they might vary under different systems of management.

Feed costs account for around 60 per cent of the cost of producing a pig, and environment is usually regarded as an important influence upon feed consumption. One might expect, therefore, that feed costs would be substantially lower in controlled-environment housing. However, in practice the thermal environment of a pig in a less-intensive house is unlikely to be significantly poorer than that of an intensively housed animal. Where a good straw bed is provided this is equivalent to an increase in air temperature of perhaps 5°C for sows, and more for growing pigs. Grouped pigs will huddle together when cold in order to reduce their heat loss, behaviour which pigs in stalls are denied. And less-intensive housing can incorporate a well-insulated kennel as a lying area, with a small air-space and a low ventilation rate in order to conserve heat. As a result, differences in feed consumption due to temperature need not be large between systems.

Capital costs were found to be fairly similar for all indoor systems and remarkably little affected by weaning age. Although more intensive systems tend to have a higher cost per square metre of floor area, they economise by stocking more densely and having a higher throughput. However, it is possible to achieve major savings by keeping pigs outdoors, thereby reducing the capital cost per pig reared to about one third of indoor systems, even allowing for more rapid depreciation of the outdoor housing.

Labour costs. It was difficult to draw very precise conclusions about labour costs which seem to vary as much between farms using similar management systems as between those using different systems. Feeding and cleaning are the time-consuming tasks, often mechanised in intensive systems. However, some extensive systems, especially outdoor ones, can be economical in their use of labour, and semi-intensive indoor systems can be designed to minimise labour use. Although the more intensive systems tend to have lower labour costs per sow, in practice there is little evidence that this is a particularly important influence on overall cost differences between systems. This is not surprising as labour costs amount to less than 15 per cent of the total costs of production.

Weaning age. After examining each of the major cost components, feed, capital and labour, we looked at a number of specific aspects of intensification. It was concluded that the effect of weaning age on production costs has been exaggerated. In practice, earlier-weaning herds fail to increase numbers of pigs reared per sow by as much as is theoretically possible, largely because of poorer breeding performance. Nonetheless, over the range 3-8 weeks it does appear that for every week earlier that weaning takes place, a sow can rear approximately an extra half pig a year. This has the advantage of spreading fixed costs over more pigs; for example, a sow will eat roughly the same amount of feed however many pigs she rears. However, this is substantially offset by the extra cost of the expensive 'creep' feeds necessary for earlier weaned pigs. In consequence, average feed cost per weaner is similar for 3 and 5 week weaning, and only about 5 per cent greater at 7 weeks.

Crates versus pens. Close confinement in crates for farrowing has caused some unease amongst welfare groups, and we examined evidence comparing their performance with that of conventional farrowing pens. We concluded that crates may save up to half a pig per litter from being crushed or overlain by the sow.

Group suckling. Group-suckling is a traditional system of grouping sows and their litters on deep straw. It is losing favour, largely because it is not suited to early weaning. It has also been claimed that pigs grow less rapidly than when they are single-suckled, and sows' breeding performance is less good. However, the available evidence provides little support for this. Although piglets' weight gain may be depressed at mixing, this is normally more than offset by subsequent above average growth which allows them to 'catch up'.

Outdoor pigs. Keeping pigs outdoors, a system which had declined in popularity, has recently begun to receive attention again. Low capital costs and advances in techniques, have made it a more attactive commercial option. It is clear that outdoor herds are capable of high levels of breeding performance, perhaps only marginally below that of comparable indoor herds. Feed costs appear slightly higher outdoors, although this is offset by greater piglet weights. The very much lower capital costs make outdoor systems an attractive alternative to indoor units, especially on light, well-drained land in a mild climate, although they may also prove competitive where conditions are less favourable.

Theoretical costings. From this information, a series of comparative theoretical costings was constructed (table 25). Overall, this suggested that cost differences between systems are not very large. A comparison of the most and the least intensive indoor systems which we considered revealed a difference of about 11 per cent in annual costs, in favour of the former. To take some more specific examples: replacing 3 week by 5 week weaning would be expected to add around 3-7 per cent to costs; if tether stalls were replaced by yards for dry sows, and flat-decks by verandah rearing accommodation, this might add between 1 and 5 per cent to costs. It is unsurprising, in the light of these results, that a wide range of different systems continue to exist side by side in the pig industry.

2. Egg production

Since the war, there has been rapid intensification of egg production. The first changes were to deep-litter systems, which housed almost half the laying flock by the mid 1950s, followed by the battery cage, which is now responsible for providing about 96 per cent of production in the UK. In this book the principle systems and their management were discussed, and a comparison of costs and retail prices made.

Non-cage systems. Non-cage systems vary from traditional free-range, which consists of small, often moveable, houses with around 150 birds an acre ranging freely to systems in which the birds are housed entirely indoors. In between are straw-yard and semi-intensive systems. The term "semi-intensive" has been used here to describe a deep-litter type of house with an outdoor grass run stocked at a rate of 1000-2000 birds an acre. Eggs produced under this system can be described as 'free-range' under the rules of the Free Range Egg Association and most of the eggs sold as 'free-range' today are produced under these conditions. Straw-yards are of the modern, fully covered type demonstrated by Dr. Sainsbury. Also considered here is the aviary system, an experimental development of deep-litter which makes greater use of the vertical dimensions of the house. By introducing an extra slatted area down the centre of the building and towards the roof, house stocking density can be increased while allowing the same 'floor' area per bird as on deep-litter.

Cost comparison. From a detailed exploration of the available evidence, theoretical costings were developed for the various systems.

Retail egg prices per dozen were estimated to be as follows for 1980/81:

Battery cage	£0.70
Deep-litter	£0.73
Aviary	£0.74
Straw-yard	£0.81
Semi-intensive	£0.86
Free-range	£0.97

Lower egg yield is the main reason for the greater expense of non-cage eggs, and is due to poorer environmental control, higher mortality and a larger number of lost and second quality eggs. The cost of labour is the other major factor, becoming more important the less intensive the system. In contrast, capital costs are a significant element in the extra cost of producing eggs using the deep-litter and aviary systems, but are less important amongst the more extensive systems.

From the figures presented here there is very little difference between the retail price of battery eggs, and what one would expect from the intensive flock systems—deep litter and aviary. However, it should be recognised that there is dispute, even within the welfare movement, as to how far intensive flock systems represent an improvement in the welfare of the stock. While they undoubtedly permit more freedom and a greater range of behaviour than commercial cages, they are more vulnerable to poor management which can lead to serious 'vice' problems. The less-intensive flock systems would increase retail prices more significantly: straw-yard by about 16 per cent, semi-intensive by about 23 per cent and free-range by about 38 per cent. However, if, as seems quite possible, legislation is introduced to limit the number of birds which could be kept in a 'standard' commercial cage of around 20″x18″, battery cage production costs would rise, affecting the comparison with non-cage systems. For example, if battery cage producers were to keep only 3 birds in a standard cage, instead of the present 5, this would make straw yard and semi-intensive systems competitive, and the intensive flock systems could have lower costs than cages. This illustrates how finely balanced the comparison is, and suggests that limiting cage size may not be the most appropriate way of trying to reconcile cost and welfare considerations.

In the light of these figures, it is at first-sight surprising that 'free-range' eggs are often so expensive in the shops. To explore this a survey of 94 shops retailing non-cage eggs was undertaken during June 1980. On average, they were being sold at a price 30 per cent more than battery eggs outside London, and 45 per cent more in London.

However, the price to the retailer was only around 20 per cent greater than for battery eggs. In other words, retailers' margins are substantially higher for non-cage than for cage eggs, reflecting the tendency to market them as 'health' foods in low-turnover, high mark-up outlets.

3. Wider implications

Although this study focusses on the production costs of different methods of livestock husbandry, there are important wider implications which also need to be considered in an economic appraisal. We examined two areas: the environmental problems associated with the handling of livestock wastes; and the energy costs of different systems.

Livestock wastes: As livestock units have grown larger, their links with the surrounding land have become more tenuous and increasingly they handle waste as a liquid slurry rather than as manure. As a result, waste disposal has become a source of pollution and nuisance. Slurry has been implicated in eutrophication of inland waterways, rising nitrate levels in water supplies, the spread of disease organisms (particularly Salmonella), damage to the fertility of agricultural land, and smell nuisance. At present, very little slurry is treated before disposal, but increased public concern may force intensive livestock farmers to take these problems more seriously. If producers are required to satisfy more rigorous standards, the result is likely to be either more sophisticated, and costly, treatment of slurry or a resurgence of interest in traditional systems. This will greatly reduce the cost advantage of existing slurry systems, and thus affect the competitiveness of those intensive units which use them.

Energy use. It has been suggested that one implication of intensification in livestock farming has been the substitution of energy for labour through mechanisation. If intensification has led to reduced energy efficiency, there might be an argument for reversing present trends as energy becomes more expensive in the future. However, much less energy is used directly in livestock rearing than in the production and processing of animal feed. Controlled environment housing relies mainly upon insulation and high stocking densities to maintain optimal temperatures. Relatively little energy is needed for supplementary heating, ventilation and other mechanical functions. On balance, energy efficiency is unlikely to be a significant reason for choosing between systems.

4. Changing systems

While some less-intensive systems, outdoor pig production for example, might be expected to gain in popularity, the central position of intensive systems is unlikely to be assailed by market forces alone in the near future. This position is reinforced by the inadequacy of information reaching consumers, and the shrinking profit margins per animal within which producers have to operate. However, the possibility of farm animal welfare legislation which would alter this balance is widely acknowledged. Some implications of a widespread change to less-intensive systems were briefly considered.

Land availability. It is sometimes claimed that less-intensive systems are no longer a feasible option because land is too valuable a resource to be used extensively. This anxiety is difficult to substantiate. The most extensive systems, such as free-range poultry are compatible with other land uses. Although they require a large area of land, poultry make little claim on its productivity because birds can be run alongside other stock. Even semi-intensive egg production, requiring exclusive use of land stocked at around 2,000 birds per acre, would require less land than is currently being lost annually to urban development. Furthermore, there would be ample opportunities during part of the year, for using pigs and poultry which are run outdoors as a break crop or a catch crop, where the opportunity cost would be low.

Straw. A further difficulty sometimes cited is that less-intensive systems would require more straw for bedding than is likely to be available. However, between a quarter and a half of all cereal straw is currently burned. Some farmers claim that this is an important practice designed to 'clean' the land by reducing weed seeds and disease pathogens and disposing of stubble. However, the evidence suggests that straw burning is more important as a cheap and convenient way of disposing of surplus straw than as a way of enhancing subsequent crop yields. It appears that fairly small increases in price would encourage farmers to make available for other uses much of the straw which is presently burned.

Producers It is argued that regulations prohibiting the use of certain systems need not damage the interests of farmers as a whole even if they do add to production costs. Some farmers, those who had invested in techniques subsequently prohibited, or those unable to adapt to changed circumstances, would suffer, although provision for compensation would alleviate this. The industry as a whole, however, would remain profitable as, under competitive conditions, it is the consumer who would be forced to pay if production costs rose. An important proviso is that welfare restrictions are not applied only to one group of farmers— UK producers, for example. Ideally, a common EEC approach would be agreed. Otherwise, any constraints which affected competitiveness and were introduced unilaterally in the UK would need to be implemented alongside measures designed to protect British farmers.

This is contrary to the spirit of the Common Agricultural Policy, although in practice examples of unilateral discrimination abound in the form of subsidies, grants and health regulations.

Consumers. On the whole, consumers can be expected to resist increases in the price of food. However, there is evidence that, increasingly, consumers are concerned with the quality of their purchases, particularly when incomes are rising. Consumers associate 'quality' to some extent with the system under which animal products are produced, hence the 'traditional', 'natural', 'farm fresh' images projected by those marketing meat and eggs. Furthermore, although there is little precise information available, concern about animal welfare undoubtedly influences some consumers' behaviour.

Transition costs. Quite apart from any effect upon the cost of production, any change from one system to another will result in additional investment and expenditure in the short term. These transition costs are the result of the premature obsolescence of skills, experience and equipment. It is difficult to make any estimate of what these costs might be as many intensive buildings would have some alternative uses, and the remaining economic life of equipment threatened with obsolescence is uncertain. Much of the existing stock of battery cages was installed 10 or more years ago and is now reaching the end of its life. In the pig industry, however, more intensive methods are still being newly adopted, and therefore much of the equipment now in use is still fairly new. Transition time is a key factor; costs would be lower the longer the period allowed for change, as this would minimise disruption and shorten the outstanding economic life of capital rendered prematurely obsolete.

To conclude, there is no neat, clear-cut relationship between intensity and cost in egg production and breeding pigs. There can be no doubt that economic forces have led to the development of intensive systems which have proved very competitive. However, the same economic forces have also been shaping the development of less-intensive systems with the result that the economic advantages of intensive systems are not as marked as is often claimed or supposed. This is certainly the case with pigs where there are a number of competitive alternatives to intensive systems, which explains the co-existance of a wide range of husbandry methods. Moving away from certain intensive practices, such as early-weaning, strawless systems, or the use of stalls, would have very little effect upon production costs. In egg production, the advantages of battery cages are small in relation to intensive indoor flock systems. A commercial niche exists even for the more extensive producers, as demand for free-range eggs is growing, despite their price.

At the same time, it is evident that major changes are unlikely to come about without new welfare legislation.

Frequently such legislation is resisted on the grounds that it would lead to an unacceptable increase in costs. On the evidence presented here, this argument should be treated with scepticism. There are practical alternatives to factory farming which could be made to prosper if intensive methods were outlawed.

Appendix 1

Types of pig housing

The main distinguishing features of the different types of housing are the form of ventilation, the amount of protection from the elements, and the system of waste handling.

1. Ventilation

Ventilation methods are generally divided into 'natural' and 'mechanical' systems. Mechanical ventilation is fan-assisted, with fans either drawing air into an enclosed building or extracting air from the building, or sometimes both. The fan-speed can be controlled manually, or automatically by linking with a thermostat or a time switch. Natural ventilation makes use of the fact that warm air rises, so that air can be made to flow if there is an outlet at or near the top of a building and above the height of air inlets. Natural ventilation can be controlled by altering the size of the air inlets and/or the outlets. This is traditionally carried out manually, although automatic control of natural ventilation is the subject of experimental work underway in Scotland. (Bruce, 1979; Lorusso, 1980).

Good control over ventilation is important for pigs. In summer, pig housing must be kept cool to avoid loss of appetite and indiscriminate mucking, which creates problems for cleaning out. In winter, excessive ventilation should be avoided so that heat is not lost unnecessarily and draughts, which lead to chilling and poor performance, are not created. But ventilation must be sufficient to prevent an excessive build-up of disease agents and noxious gases in the atmosphere and to avoid condensation. This difference in summer and winter ventilation requirements, which may differ by a factor of 10, needs to be accommodated in the design of the system.

The principle advantage of mechanical ventilation is that, providing there are no power failures and the design is adequate, its effectiveness is considerably less dependent upon external climatic conditions than natural ventilation. On a hot, still day natural ventilation may be inadequate and in open-fronted buildings ventilation may be excessive on cold, windy days. In practice, efficient natural ventilation also depends upon the skill and diligence of the stockman to a greater extent than automatically controlled fan-assisted ventilation, although this will be less important if automatic controls are extended to naturally ventilated systems.

2. Environmental protection

Environmental protection can vary greatly. In a traditional outdoor system, the only shelter provided might be a half-round corrugated iron hut with straw bales or wooden boarding at either end and straw bedding. In contrast, there are buildings which are wholly enclosed, well-insulated and perhaps heated. In practice, supplementary heating is generally unnecessary and prohibitively expensive except for very young early-weaned pigs, which are particularly vulnerable to chilling.

Between these extremes lie variations on yards and kennels. A general purpose structure can house large groups of dry sows or weaners in yards, deep-straw bedding offsetting the poor environmental control resulting from low-stocking density, large air space and comparatively poor insulation. The addition of a false roof to provide straw storage, has the advantage of providing a smaller and better insulated air space. (Figure A1).

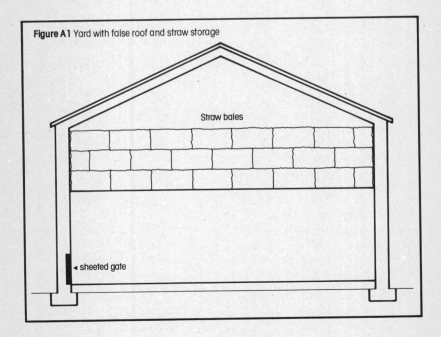

Figure A1 Yard with false roof and straw storage

Straw bales

◄ sheeted gate

The principle of a kennel building is separation of the lying area from the exercise and dunging area. The lying area, or kennel, is comparatively well-insulated and ideally kept densely stocked to maintain its temperature and to discourage mucking. Straw may be restricted to the kennel or used throughout. Depending upon climatic conditions and capital costs, the yard area can be either outdoors or under cover. (Figures A2 and A3).

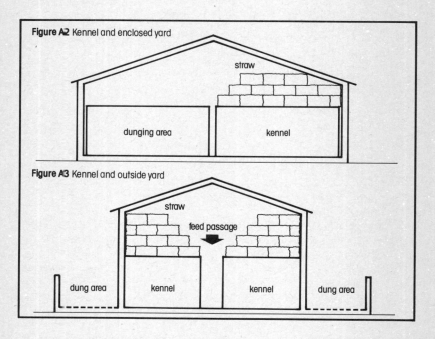

Figure A2 Kennel and enclosed yard

straw

dunging area

kennel

Figure A3 Kennel and outside yard

straw

feed passage

dung area

kennel

kennel

dung area

The same principle may be applied in the form of monopitch housing. Again, the extent of protection can vary. (Figures A4 and A5).

3. Waste handling

Waste handling systems vary according to the extent to which wastes are handled in liquid or solid form. Where wastes are allowed to accumulate, and fresh straw is added frequently to keep the pigs clean and dry the wastes are handled in solid form as farmyard manure. Where there is a separate solid-floored dunging area, this is generally scraped daily using a tractor and scraper or a hand scraper which may be mechanically driven. Pig buildings incorporating a dunging passage

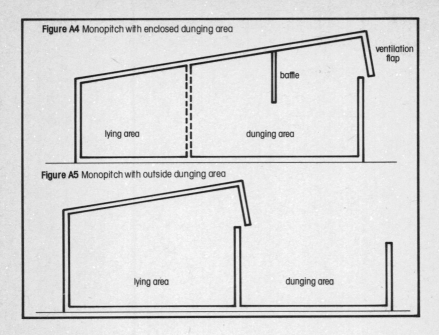

Figure A4 Monopitch with enclosed dunging area

ventilation flap

baffle

lying area

dunging area

Figure A5 Monopitch with outside dunging area

lying area

dunging area

are normally designed so that it runs the whole length of the building and pigs can be excluded from it when it is being cleaned (see Figure A6). In this way, a whole building can be cleaned very rapidly.

The waste is then scraped to a midden for storage. It may be fairly solid as it is likely to contain some straw from the bedded kennels and the liquid is often allowed to drain from the midden to a separate urine tank. The solid and liquid parts can then each be handled quite easily.

A development of this system is for the dunging areas (or, in some cases, the whole area occupied by the pig) to have slatted floors. Faeces and urine fall into a slurry channel beneath the floor (see Figure A7) where they accumulate before being transferred to a storage lagoon, or directly spread on the land by tanker or irrigation system. This saves the daily labour of cleaning the dunging areas. In most cases, little, if any, straw is used in a slatted floor system, partly because straw can block the equipment used to handle liquid slurry.

We can see now how these principles are applied in the different kinds of buildings.

Dry sows at their most intensive are tethered or enclosed in individual

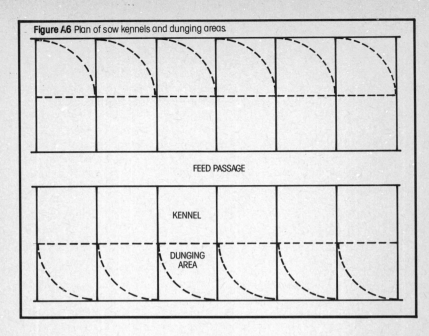

Figure A6 Plan of sow kennels and dunging areas.

FEED PASSAGE

KENNEL

DUNGING
AREA

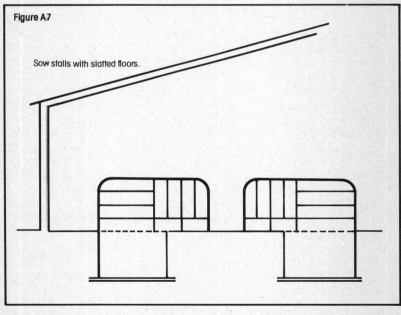

Figure A7

Sow stalls with slatted floors.

stalls. Typically, there are two or four rows with sows facing each other across a feeding passage. In full stalls, sows are enclosed on all sides by a tubular steel framework, approximately 2.1 metres by 0.6 metres. Where sows are tethered, most of the tubular frame is unnecessary, with only a metal division retained to prevent bullying. In this way space is saved and capital costs reduced. Housing needs to have a very closely controlled environment as the restriction on sow movements and the absence of bedding makes them vulnerable to conditions of adverse temperature or ventilation. Part-slatted floors are widely used for stalls, although solid floors are generally used with tie stalls. The advantages of stalls are that they allow control over individual feed intake, eliminate bullying, and are suited to mechanisation of feeding. Close observation and individual attention are comparatively easy, although lack of movement may obscure the presence of problems. The high stocking density makes it easier to maintain optimal temperatures.

Alternatively, dry sows may be group-housed. The close control of the feed consumption of individual sows which has become desirable with the increase in feed costs may be achieved by providing individual feeders. These are similar to stalls and are used only during feeding. (Figure A9). This does add significantly to capital costs, depending upon how much of the dunging and feeding area is covered, and bullying can still be a problem outside feeding times. In an effort to overcome these problems, sow cubicles have been developed. (See Figure A8). Sows lie in individual cubicles, in which they are also fed, and have access to a scraped dunging and exercise area.

Farrowing accommodation is now dominated by the farrowing crate, a tubular metal construction similar to the dry sow stall but slightly larger. (Figure A10). Its purpose is to restrain the movements of the sow so as to reduce piglet mortality due to crushing and savaging which can be a serious problem, particularly immediately after farrowing. The crate is placed within a pen where the piglets feed, sleep, exercise and dung. Straw bedding may be provided, or the floor can be partly or fully slatted. There is normally a separate 'creep' area for piglets to lie in. Creep areas are often heated to between 21° and 27°C, the temperature range within which the piglets are most comfortable but which would be unnecessarily warm for the sow. Sows may remain in crates for anything from 3-4 days to 3-4 weeks. An early move to less restrictive suckling accommodation can be associated with a secondary peak in mortality, probably due to the piglets' lack of familiarity with their new surroundings.

It is possible to design pens in which the sow is not always confined to

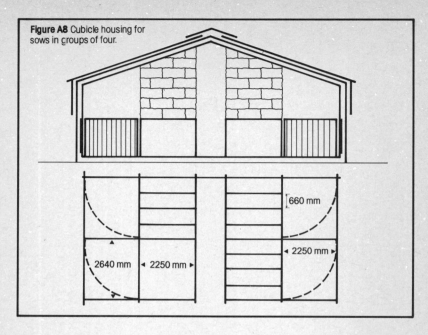

Figure A8 Cubicle housing for sows in groups of four.

660 mm

2250 mm

2640 mm ◄ 2250 mm ►

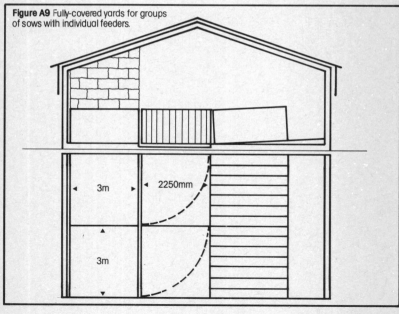

Figure A9 Fully-covered yards for groups of sows with individual feeders.

◄ 3m ► ◄ 2250mm

3m

a crate. Access to a separate dunging/exercise area may be provided twice a day for feeding and drinking. This requires extra labour, but provides the opportunity for regular inspection by the stockman while the sow is moving around. It does entail a lower stocking density, which implies higher capital cost and greater difficulty in maintaining temperatures. (Figure A11).

Traditional farrowing accommodation makes no use of crates. Farrowing rails prevent sows from crushing their young against the walls of the pen, a particular hazard, and piglets are encouraged to lie away from the sow in a creep area, where additional food and supplementary heat may be provided. The type illustrated in Figure A12 has the advantage of allowing the stockman easy access to piglets in the creep area, unlike the solari-type pen illustrates in Figure A13. The solari-type pen endeavours, not always with complete success, to maintain the house temperature by minimising the air space and using an insulated roof. The sow uses the area between the rails voluntarily and there is additional space for dunging and exercise. One advantage of the more traditional pens is that, at weaning, the sow can be taken from the litter, which can remain in the pen during the rearing and (in the case of the solari-type) finishing stages. This reduces the need for traumatic moves to new accommodation which can retard performance.

4. Rearing accommodation

Rearing accommodation can take various forms, depending particularly upon the age at which piglets are weaned. Pigs weaned between 5 and 8 weeks are generally suckled either in the traditional pens illustrated in Figures A12 and A13 or they are grouped with other sows and litters at 10-20 days in multi-suckling accommodation. Where they are single-suckled, a farrowing crate might be used in the early stages, after which it can be removed to allow the sow more freedom. Multi-suckling accommodation might be in a general purpose building, of the type shown in Figures A1 and A2, or a monopitch building of the kind illustrated in Figure A14. In either case, a creep area would need to be provided, where creep feeds* can be offered without being taken by the sow. Normally, groups of 2-6 sows are kept together, with a recommended total floor area of 60-75 square feet per sow (5.4-6.8m^2).

Pigs weaned earlier than 3 weeks are highly vulnerable and require very specialised buildings which provide a carefully controlled environment and minimal contact with their dung. Some piglets are kept in wire cages, in 2 or 3 tiers, with dung passing onto a solid surface beneath each cage from which it is regularly scraped. Commercial use

* highly concentrated milk-substitute feeds designed to encourage weaning onto solid food.

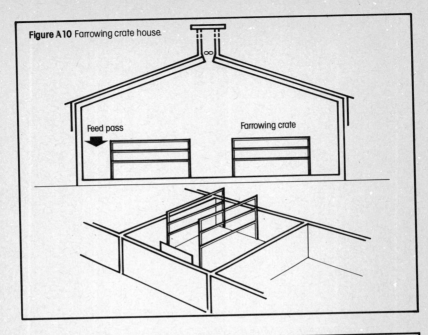

Figure A10 Farrowing crate house.

Feed pass

Farrowing crate

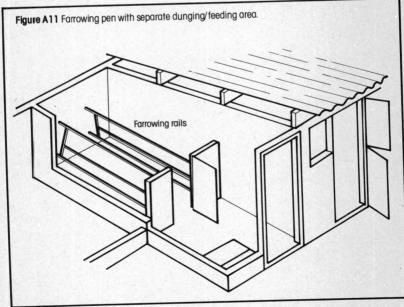

Figure A11 Farrowing pen with separate dunging/feeding area.

Farrowing rails

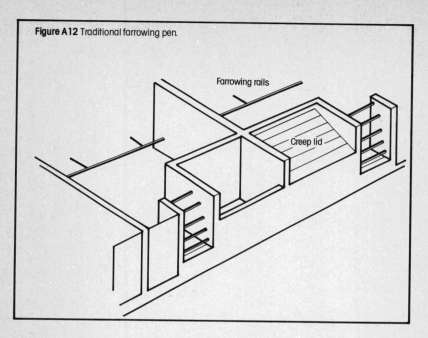

Figure A12 Traditional farrowing pen.

Farrowing rails

Creep lid

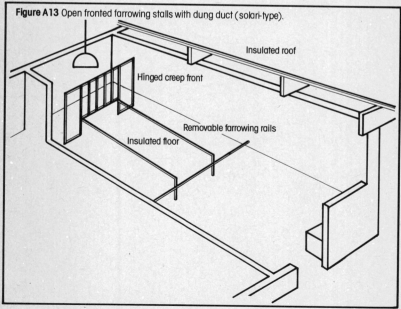

Figure A13 Open fronted farrowing stalls with dung duct (solari-type).

Insulated roof

Hinged creep front

Removable farrowing rails

Insulated floor

120

of cages is very limited and, in general, no advantage over 3-4 weeks weaning has been established.

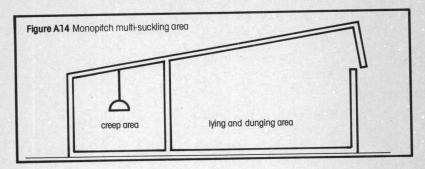

Figure A14 Monopitch multi-suckling area

creep area

lying and dunging area

For weaning at 3-5 weeks, the choice is between flat-decks and verandahs. Flat decks can be loosely described as single-tier cages. Typically the whole floor is of slatted metal, with open metal pen divisions. Each cage holds 10-30 weaners and there might be 6-12 cages in a room, a number of rooms forming a house. The house is totally enclosed, well insulated, and there is close control of the environment, including heating in the early stages. Pigs may remain in flat-decks for anything up to 12 weeks, periodically thinned out as they grow, although the comparatively high cost of this form of accommodation encourages transfer to cheaper accommodation as soon as the pigs are sufficiently hardy. Verandahs consist of a kennelled lying area with a separate open dung area. The kennel is well insulated to conserve the heat given off by the stock, although for pigs under 5 weeks, particularly where straw is not used, supplementary heat may be provided. The dunging area can be slatted or solid and may be either enclosed within a

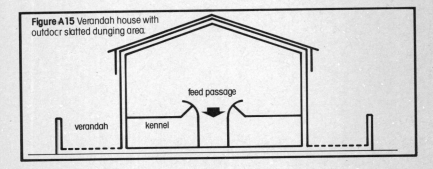

Figure A15 Verandah house with outdoor slatted dunging area.

feed passage

verandah   kennel

building, especially for early weaned pigs, or outdoors so as to reduce costs and provide access to fresh air. (Figure A15).

Both kennels and flat-decks ensure good separation of pigs and muck and provide a warm environment. Although pigs in verandahs have more room in total than those in flat-decks, they are packed very tightly in the kennel area to ensure sufficient warmth. This can make inspection difficult. Close control of the environment is also more difficult than in a flat-deck, although usually not dependent upon supplementary heating. In verandahs there are generally fewer pigs, usually 15-30 within an air-space, which inhibits the spread of air-borne diseases.

Pigs of 5-12 weeks may be kept in weaner pools, which are large, deep straw-yards, naturally ventilated, and often with a kennelled sleeping area. Group size is normally large, 30-60 or more. Monopitch housing may also be used, often on a straw based system.

Appendix 2

Some notes on thermal environment and feed consumption

1. Temperature difference between the inside and the outside of buildings

An expression for this is as follows:

$$T = Hm/1200V + C$$

where: Hm = metabolic heat given off, watts per animal

V = ventilation rate, m^3/second per animal

C = conductance loss, i.e. heat lost per animal per degree temperature difference between inside and outside, watts/°C per animal

and $C = A$ (U floor + U roof + 2H/WU wall)

where: A = total building floor area per animal

H = height of building to eaves, m

W = width of building, m

U = thermal conductivity ('U' value), watts/m^2 per °C

For pigs the following assumptions are made:

minimum ventilation rate is $0.375m^3$/h per kg

Hm = 220 watts for dry sows

310 watts for sow and litter

60 watts for a 30kg weaner

For the sow and litter, the contribution to total building heat made by localised creep heaters is not considered.

An insulation standard (following Bruce (1974)[1] of $C = 1.75A$ is assumed, as follows:

U roof = 0.73

U walls = 1.19

U floor = 0.50 (no straw)

Assume that $H = 2.2m$ and $W = 10m$.

For poultry the following assumptions are made:

Minimum ventilation rate of $11m^3$/hour per kg of feed per day, equivalent to approximately $1.3m^3$/hour per bird per day.

Although heat output is related to temperature, we simplify and assume that it is constant at 11.3 watts/bird.

U values are assumed as follows:

U roof	= 0.6
U walls	= 1.1
	= 0.5, except for litter systems in which it is assumed to be zero.

2. Developing estimates of feed consumption of layers under different systems of management

The results presented as Table 31 take account of thermal environment and egg yield on feed consumption. Emmans (1974)[2] examined the results of 14 experiments and concluded that for each degree drop between -3°C and +15°C birds will consume 1.85 calories as extra feed, and from +15°C to 29°C they will consume an extra 3.93 calories. These are clearly intended as approximations as presumably the relationship is a continuous one.

Hewson (1979)[3] uses this data to estimate feed consumption amongst extensively kept layers. Calculations on this basis need to be treated with some scepticism because air temperature is only one component of a layer's thermal environment. Air speed and radiated heat are important as well and differ between management types. Unfortunately the difficulty of measuring these effects leads to air temperature being adopted as a proxy. Even the experimental data on which this relationship is based needs cautious interpretation. Apart from difficulties in experimental design, the measurements are principally for caged stock kept at temperatures of 10°C and more. Whether observed relationships persist below this level is uncertain.

Hewson assumes that under extensive systems, the thermal environment can be approximated by external ambient temperatures. However, under all systems birds spend a substantial part of their time in some form of housing which is designed to improve their environment. Birds are completely housed at least between sunset and sunrise— the coldest part of the day— and may be outdoors for 4 hours or less each day during the coldest months. Furthermore, even where there is free access to range, birds will choose to spend some time indoors (nesting, for example), and this is likely to be affected by weather conditions. The internal temperature will clearly vary according to the kind of housing, depending upon insulation, air-space per bird, ventilation, availability and quality of litter, and conditions outdoors.

One complicating factor in considering litter systems is that the floor is both a source of heat gain to the building through bacterial action, as well as a source of heat loss. It can be assumed, arbitrarily but conservatively, that net heat flow through the floor is zero.

Table A THE EFFECT OF STOCKING DENSITY ON INTERNAL—EXTERNAL TEMPERATURE GRADIENT UNDER DEEP-LITTER CONDITIONS

ft² per bird	m² per bird	birds/m²	temperature gradient (°C)
*1.0	0.093	10.75	21.4
1.5	0.139	7.19	19.6
2.0	0.186	5.38	18.1
2.5	0.232	4.31	16.8
3.0	0.279	3.58	15.6

* obtainable under aviary conditions.

Temperature gradients in excess of 20°C are normal under cage conditions, depending upon layout and cage stocking density. Birds kept in extensive systems will spend part of their day out of doors, at the ambient temperature. Furthermore, pop-holes or doorways to the outside will make it more difficult to achieve minimum ventilation rates. Consequently, in estimates it is assumed that the temperature gradients for semi-intensive and free-range housing are 10°C and 5°C respectively. Additional assumptions are: birds kept semi-intensively spend 15 per cent, and range birds spend 40 per cent, of their time outdoors. In practice, the coldest part of the day will be spent indoors, and birds will be out less during the colder months when there is less daylight, but for simplicity this is ignored.

Egg-production is assumed to differ between systems, and this will affect feed intake. Wills (1975)[4] draws upon estimates of factors determining feed requirements to suggest that each gram of egg requires between 7.78 and 11.25 kJ of ME intake. This is approximately 0.4-0.6 kgs of feed per 10 eggs.

Combining the effect of temperature and egg yield suggests the estimates presented in Table 31.

References

1 Introduction
1. House of Commons Agriculture Committee, *Animal Welfare in Poultry, Pig & Veal Calf Production,* First Report Session 1980-81, HMSO, 1981.
2. MAFF, *Animal Welfare in Poultry Pig & Veal Calf Production,* Cmmd 8451, HMSO, Dec. 1981.
3. Burton, D.A., *The Economics of Egg Production,* Dept. of Agricultural Economics, University of Manchester, 1978.
4. Thomas, W.J.K., Some Aspects of Pig Production, 1953/4-1978/9 in Thomas, W.J.K. (ed) *Pig Production,* University of Exeter, Ag. Ecs. Unit, Report No 209, 1980.
5. MAFF, *Egg Yield Survey,* Press Notices 1971-80.
6. Meat & Livestock Commission, Hit Count data, Mimeo, 1972-80.
7. Richardson, D.I.S., *Economic Progress & Problems of the Egg and Poultrymeat Industries,* University of Manchester, Dept. of Ag. Ecs., 1978.
8. Thomas, W.J.K., 1980, *op. cit.*

2 Breeding pigs
1. MAFF, *Housing the Pig,* Bulletin 160, HMSO 1972.
2. Bruce, J.M., Heating requirements for pig housing, *Farm Buildings Progress, 36,* April 1974.
3. Sainsbury D. Sainsbury S., *Livestock Health & Housing,* Balliere Tindall, 1979.
4. Holmes, C.W. & Close, W.H., The influence of climatic variables on energy metabolism and associated aspects of productivity in the pig, in Haresign, W., Swan, H. & Lewis, D., *Nutrition & the Climatic Environment,* Butterworth, 1977.
5. Bruce, J.M. & Clark, J.J., Models of Meat production and critical temperature for growing pigs, *Animal Production, 28,* 1979.
6. Armstrong, B., A technique for assessing the economics of environmental control in farm buildings, in Monteith, J.L. & Mount, L.E. (Eds) *Heat Loss from Animals & Man,* University of Nottingham Twentieth Easter School in Agricultural Science, Butterworths, 1973.
7. Smith, A.T., Effect on animal performance by the environment, paper to conference on intensive livestock systems at National Agricultural Centre, January 1981.
8. Ibid.
9. Stephens, D.B., The metabolic rates of newborn pigs in relation to floor insulation and ambient temperature, *Animal Production 13,* 1971.
10. Verstegen, M.W.A. & Hel, W. van der, The effects of temperature & type of floor on metabolic rate & effective critical temperature in groups of growing pigs, *Animal Production 18,* 1974.
11. Moustegard, J., Nielson, P.B. & Sorensen, P.H., Influence of straw bedding & huddling on daily gain & feed conversion in the weight ranges 40-90 kg at an ambient temperature of 3°C, *Royal Vet. Agric. College Sterility Research Instit., Annual Rpt. 173.* 1959.
12. Op. cit.

126

13. Bruce, J.M., Heat loss from animals to floors, *Farm Buildings Progress 55,* Jan 1979.
14. *Op. cit.*
15. *Op. cit.*
16. Sainsbury, D. pers. comm.
17. Thornton, K., *Practical Pig Production,* 2nd edn. Farming Press, 1978.
18. Robertson, A.M. & Kelly, M., Flat-deck accommodation for pigs, *Farm Building Progress,* Oct. 1978.
19. Hafez, E.S.E., *The Behaviour of Domestic Animals,* 3rd edn., 1978.
20. National Agricultural Centre, Pig Demonstration Unit, Report, 1979.
21. Holsten, E.E., *Piglet performance in weaner houses,* MAFF, 1978.
22. Centre for Agricultural Strategy, *Capital for Agriculture,* CAS Report No 5, 1978.
23. UK Agricultural Departments, *Animal Welfare in Poultry, Pig & Veal Calf Production,* Minutes of Evidence to House of Commons Agriculture Committee, HMSO, November 1980.
24. Nix J., *Farm Management Pocketbook,* 10th edn., Wye Coll., 1979.
25. Meat & Livestock Commission, *Pig Facts,* 1979.
26. ADAS, *Cost of Buildings,* 1980.
27. Agro-Business Consultants Ltd., *The Agricultural Budgeting & Costing Book No 10,* 1980.
28. Baxter-Parker Ltd, Equipment required for a 330 sow extensive unit, unpublished mimeo, 1980.
29. Wight, H.J. & Clark, J.J., *Farm Building Cost Guide 1980,* SFBIU, 1980.
30. Clark J.J. & Bruce J.M., Estimating sow herd accommodation, *Farm Building Progress,* April 1979.
31. Thomas, W.J.K., Some aspects of pig production — 1953/4-1978/9 in Thomas W.J.K. (ed) *Pig Production,* University of Exeter, AG, Ecs. Unit, Report No 209, 1980.
32. *Op. cit.*
33. Daniel, R., Kadlec, J.E., Morris, W.H., Jones H.W., Hinkle, C.N., Conrad, J.H., & Dale, A.C., Productivity & cost of swine farrowing & nursery systems. *Res. Prog. Rep. Univ. Purdue,* 315, 1967.
34. Ridgeon, R.F., *Pig Management Scheme Results for 1980,* University of Cambridge, Dept. of Land Economy, 1980.
35. Meat and Livestock Commission, Sow productivity, *Pig Improvement Services Newsletter No 14,* March 1980.
36. Ridgeon, R.F., *Pig Management Scheme Results for 1979,* University of Cambridge, Dept. of Land Economy, 1979.
37. Meat & Livestock Commission, 1980, op. cit.
38. Brake, J.H.A. te, An assessment of the most profitable length of lactation for producing piglets of 20 kg body weight, *Livestock Production Science 5,* 1978.
39. Ridgeon, R.F., op. cit, 1978, 1979, 1980.
40. *Op. cit.*
41. Meat & Livestock Commission, *Commercial Pig Production Yearbook 1979,* MLC 1980.
42. Meat & Livestock Commission, *Commercial Pig Production Yearbook 1980,* MLC, 1981.
43. Ridgeon, R.F., 1978, 1979, 1980, *op. cit.*
44. Brake, J.H.A. te, 1978, *op. cit.*
45. Meat & Livestock Commission, 1979, *Op. cit.*
46. Ridgeon, R.F., 1980, *op. cit.*
47. Sturrock, H., Cutting suckling losses, *Pig Farming,* supplement, November 1977.
48. Baxter, M.R. & Petherick, J.C., The effect of restraint on parturition in the sow, paper to Pig. Vet. Soc. Denmark, June 1980.

49. Robertson, J.B., Laird, R., Hall, J.K.S., Forsyth, R.J., Thomson, J.M., & Waller-Love, J., A comparison of 2 indoor farrowing systems for sows, *Animal Production, 8,* 1966.
50. Robertson, A.M., Accommodation for farrowing & lactating sows, *Farm Building Progress,* April 1977.
51. Bauman, R.H., Kadlec, J.E., & Powlen, P.A., Some factors affecting death loss in baby pigs, *Purdue University Research Bulletin No 810,* June 1966.
52. Seidel, C., Cost savings of farrowing containers, Baxter-Parker Ltd., unpublished mimeo, 1980.
53. Parrish, D., Managing the large pig unit, *Big Farm Management,* Oct., Nov & Dec 1972.
54. Boddington, M.A.B., *Outdoor Pig Production: report on an economic investigation,* Wye Coll., 1971.
55. Meat & Livestock Commission, 1980 & 1981, *Op. cit.*
56. Newland, P.I., *Pig Management Survey,* unpublished mimeo, NFU, 1980.
57. Thornton, 1978, *Op. cit.*
58. Hillyer, M., Suckling sows in groups, *Pig Farming,* June 1976.
59. MAFF, 1972, *Op. cit.*
60. MAFF, *Pig Husbandry & Management,* 3rd edition, HMSO, 1977.
61. Hillyer, 1976, *Op. cit.*
62. Petchey, A.M., Dodsworth, T.L. & English, P.R., The performance of sows & litters penned individually or grouped in late lactation, *Animal Production 27,* 1978.
63. Thornton, 1978, *Op. cit.*
64. Royal Agricultural Society of England, *Fream's Elements of Agriculture,* 14th edn, 1962.
65. Boddington, 1971, *op. cit.*
66. Ibid.
67. Meat & Livestock Commission, Pig Improvement Services Regional H I T Count; Breeding Herds, mimeo, MLC March 1979.
68. Newland, 1980, *op. cit.*
69. Shepherd, A, Out for a start, *Pig Farming,* Dec 1980.
70. Hope, H., At home on the range, *Farmers Weekly,* Nov 21, 1980.
71. Harvey, G., outdoor herds on the increase, *Farmers Weekly,* Nov 21, 1980.
72. Mitchell, W.S., An outdoor breeding herd, *Just Pigs 13,* 1980.
73. Roach, B., Hewitt, C., & Foster, J., Come-back for outdoor pig keeping? *Pig Farming,* July 1981.
74. *Op. cit.*
75. *Op. cit.*
76. Meat & Livestock Commission, *Commercial Pig Production Yearbook 1979,* MLC 1980.

3 Egg production

1. Orton, C.R., Surveys of egg yields in England & Wales, 1948-69, *British Poultry Science 12,* 1971.
2. Blount, W.P., *Intensive Livestock Farming,* Heinemanne, 1968.
3. Morris, T.R., The influence of photoperiod on reproduction in farm animals, University of Nottingham 31st Easter School in Agricultural Science, April 1980.
4. Charles, D.R., Environment for poultry, *Veterinary Record,* April 5, 1980.
5. Gowe, R.S., A comparison of the egg production of 7 S C White Leghorn strains housed in laying batteries and floor pens, *Poultry Science 35,* 1956.
6. Lowry, D.C., Lerner I.M., & Taylor, L.W. Intra-flock genetic merit under floor & cage management, *Poultry Science 3,* 1959.

128

7. Bailey, B.B., Quisenberry, J.H., & Taylor, J., A comparison of performance of layers in cage & floor housing, *Poultry Science 38*, 1959.
8. Logan. V.A., Influences of cage versus floor, density & dubbing on laying house performance, *Poultry Science 44*, 1965.
9. Plumart, P.E., Muller, R.D., & Carlson, C.W., A comparison of slat floors, litter floors & cages for laying hens, *Second Annual Poultry Field Day* Dept. of Animal Science, South Dakota State University, 1970.
10. Bareham, J.R., Effects of cages & semi-intensive deep-litter pens on the behaviour, adrenal response & production in 2 strains of laying hens, *British Veterinary Journal, 128*, 1972.
11. Swiss Foundation for the Promotion of Poultry Breeding & Keeping, Final Report on the performance trial for laying animals 1976/78, 1978.
12. Hale, R.W., Genotype-environment interactions in a comparison of the cage & semi-intensive systems for laying hens, *British Poultry Science 2*, 1961.
13. Coles, R., The influence of housing systems & ageing of layers on egg quality & hatching, *Poultry Science 39*, 1960.
14. Wegner, R.M., Experiments on the production & behaviour of poultry 1977, *Deutsche Geflugelwirtschaft und Schweine*, April 1978, reproduced in Elson, H.A., Report of a visit to West Germany, MAFF, 1978.
15. Elson, 1978, *Op. cit.*
16. Sainsbury D. & Sainsbury S., *Livestock Health & Housing*, Balliere Tindall, 1979.
17. Swiss Centre for Poultry, *Feed Costs for Laying Animals*, 1978.
18. Burton. D.A., *The Economics of Egg Production*, Dept. of Agricultural Economics, University of Manchester, 1978.
19. *Op. cit.*
20. *Op. cit.*
21. *Op. cit.*
22. *Op. cit.*
23. *Op. cit.*
24. *Op. cit.*
25. Op. cit.
26. Elson, 1978, *Op. cit.*
27. Wills, R.J.R., Economic components of egg production in Freeman, B.M., & Boorman, K.N., *Economic Factors Affecting Egg Production*, 1975.
28. Hewson, P.F.S., Extensive systems—some facts and figures, paper to conference at Plumpton Agric. College, Feb 21, 1979.
29. University of Bristol, A comparison of certain economic factors in extensive & intensive systems of egg production, *Farm Economics Notes No. 5*, 1954.
30. Halnan, E.T., & Garner, F.H., *The Principles & Practice of Feeding Farm Animals*, 1953.
31. Blount, 1968, *Op. cit.*
32. Sainsbury & Sainsbury, 1979, *Op. cit.*
33. Elson, 1978, *Op. cit.*
34. Burton, D.A., A costings perspective, written evidence to Agriculture Committee of the House of Commons, 1981.
35. Morris, T.R., Some economic aspects of animal welfare, paper to Reading University Agricultural Club 15th Annual Conference, Feb 11, 1981.
36. Sturrock, F.G. *Planning Farm Work*, MAFF Bulletin 172, HMSO, 1960.
37. MAFF, Earnings & hours of agricultural workers in England & Wales, *Press Notice*, Oct 20, 1980.
38. Sturrock, 1960, *Op. cit.*

39. MAFF, *Intensive Poultry Management for Egg Production*, Bulletin 152, HMSO, 1976.
40. Hewson, 1979, *Op. cit.*
41. *Op. cit.*
42. ADAS Farm Animal Welfare Group, *A cost comparison of commercial egg production systems*, MAFF, 1981.
43. UK Agriculture Departments, *Animal Welfare in Poultry, Pig & Veal Calf Production*, Minutes of Evidence to House of Commons Agriculture Committee, HMSO, November 1980.
44. ADAS, 1981, *Op. cit.*
45. Burton, 1981, *Op. cit.*
46. Morris, 1981, *Op. cit.*

4   Some wider costs
1. Newland, P I, *Pig management survey*, unpublished mimeo, NFU, 1980.
2. Evans, M.R. & Owen, J.C., Factors affecting the concentration of fecal bacteria in land-drainage water, *Journal of Microbiology 71*, 1972.
3. Jones, P.W., Bew, J., & Burrows, M.R., The occurrence of salmonellas, microbacteria & pathogenic strains of E. coli in pig slurry, *Journal of Hygiene (Cambridge) 77*, 1976.
4. Evans & Owen, 1972, *Op. cit.*
5. Royal Commission on Environmental Pollution, *Agriculture & Pollution*, Seventh Report, Cmnd 7644, HMSO, 1979.
6. Hodge, I.D., On the local environmental impact of livestock production, *Journal of Agricultural Economics 27*, 1978.
7. Nielson, C, Mind that smell, *British Farmer & Stockbreeder*, Supplement, April 26, 1980.
8. Royal Commission on Environmental Pollution, 1979, *Op. cit.*
9. Nielson, 1980, *Op. cit.*
10. Gill, D., *Intensive Livestock Unit Subject Plan*, Humberside County Council, 1979.
11. White, D.J., Energy use in agricultural systems, *Agricultural Engineering 30*, 1975.
12. Wathes, C.M.., Energetic efficiencies of alternative systems of egg production, *Poultry Science Co*, 1981.
13. Thompson, D., Broilers burning fossil energy, *Poultry World*, October 23, 1980.
14. Wathes, 1981, *Op. cit.*

5   Changing systems
1. House of Commons Agriculture Committee, *Animal Welfare in Poultry, Pig & Veal Calf Production*, First Report session 1980-81, HMSO, 1981.
2. Countryside Review Committee, *Food Production in the Countryside*, Topic paper 3, HMSO, 1978.
3. Henderson, G., *The Farming Ladder*, Faber, 1944.
4. Boddington, M.A.B., *Outdoor Pig Production: Report on an Economic Investigation*, Wye College, 1971.
5. Hawkins, H., Outdoor pig rearing, in Universities Federation for Animal Welfare, *Alternatives to Intensive Husbandry Rearing*, Proceeding of a Symposium at Wye College, July 1981, UFAW, 1981.
6. Whittemore, C.T., & Elsley, F.W.H., *Practical Pig Nutrition*, 1977.
7. Boddington, 1971, *Op. cit.*
8. Ibid.
9. Meat & Livestock Commission, *Commercial Pig Production Yearbook 1979*, MLC, 1980.

10. Dean, T.W.R., *Straw Price & Availability*, Paper Industries Research Assocn, 1977.
11. Royal Commission on Environmental Pollution, *Agriculture & Pollution*, Seventh Report, Cmnd 7644, HMSO, 1979.
12. Newland, P.I., *Pig Management Survey*, unpublished mimeo, NFU 1980.
13. Hughes, R.G., Arable Farmers' problems with straw, in Grossbard, E, (ed) *Straw Decay & Its Effects on Disposal & Utilisation*, 1979.
14. Cited in Staniforth, A.R., *Cereal Straw*, OUP, 1979.
15. Ibid.
16. Elliott, J.G., Straw & reed problems, paper to NFU Conference, Straw—an asset or liability to the farmer? 3rd April, 1978.
17. Staniforth, 1979, *Op. cit.*
18. Edwards, C.A., & Lofty, J.R., The effects of soil residues & their disposal on the soil fauna, in Grossbard, 1979, *Op. cit.*
19. Staniforth, 1979, *Op. cit.*
20. Wood, R.S., Straw burning—purposes, methods, costs, in ADAS, *Report on Straw Utilisation Conference*, MAFF, 1977.
21. ADAS, *Fate of Cereal Straw*, MAFF, 1976.
22. Morris, J., Radley, R.W., Smith, D.L.O., & Plom, A., Straw supply in the UK with particular reference to the industrial usage, *Agricultural Progress 51*, 1976.

Appendix 2

1. Bruce, J.M., Heating requirements for pighousing, *Farm Buildings Progress 36*, April 1974.
2. Emmans, G.C., The effect of temperature on the laying performance of hens, in Morris, T.M. & Freeman, B.M. (eds) *Energy Requirements of Poultry*.
3. Hewson, P.F.S., Extensive systems—some facts & figures, paper to conference at Plumpton Agric, College, Feb. 21, 1979.
4. Wills, R.J.R., Economic components of egg production, in Freeman, B.M. & Boorman, K.N., *Economic Factors Affecting Egg Production, 1975*.

Notes

Notes